In-Memory Data Management Research

Series Editor

Prof. Dr. Dr. h.c. Hasso Plattner
Hasso Plattner Institute
Potsdam, Germany

For further volumes:
http://www.springer.com/series/11642

Jan Schaffner

Multi Tenancy for Cloud-Based In-Memory Column Databases

Workload Management and Data Placement

Springer

Jan Schaffner
Hasso Plattner Institute for IT Systems Engineering
Potsdam
Germany

Dissertation at the University of Potsdam, Hasso Plattner Institute for IT Systems Engineering

ISBN 978-3-319-00496-9   ISBN 978-3-319-00497-6 (eBook)
DOI 10.1007/978-3-319-00497-6
Springer Heidelberg New York Dordrecht London

Library of Congress Control Number: 2013942310

© Springer International Publishing Switzerland 2014
This work is subject to copyright. All rights are reserved by the Publisher, whether the whole or part of the material is concerned, specifically the rights of translation, reprinting, reuse of illustrations, recitation, broadcasting, reproduction on microfilms or in any other physical way, and transmission or information storage and retrieval, electronic adaptation, computer software, or by similar or dissimilar methodology now known or hereafter developed. Exempted from this legal reservation are brief excerpts in connection with reviews or scholarly analysis or material supplied specifically for the purpose of being entered and executed on a computer system, for exclusive use by the purchaser of the work. Duplication of this publication or parts thereof is permitted only under the provisions of the Copyright Law of the Publisher's location, in its current version, and permission for use must always be obtained from Springer. Permissions for use may be obtained through RightsLink at the Copyright Clearance Center. Violations are liable to prosecution under the respective Copyright Law.
The use of general descriptive names, registered names, trademarks, service marks, etc. in this publication does not imply, even in the absence of a specific statement, that such names are exempt from the relevant protective laws and regulations and therefore free for general use.
While the advice and information in this book are believed to be true and accurate at the date of publication, neither the authors nor the editors nor the publisher can accept any legal responsibility for any errors or omissions that may be made. The publisher makes no warranty, express or implied, with respect to the material contained herein.

Printed on acid-free paper

Springer is part of Springer Science+Business Media (www.springer.com)

To Dean B. Jacobs (1958–2013), a mentor and friend.

Acknowledgments

I would like to thank my advisor, Prof. Hasso Plattner, for a six year long learning opportunity. Hasso is perhaps the most vibrant and enthusiastic person I have ever come across. If it were not for him and his energy, the HANA project and also this thesis would not have happened.

I would also like to thank my wonderful colleagues at HPI, most notably the ones involved in the HANA project: Anja Bog (now with SAP), Martin Faust, Martin Grund (now with the University of Fribourg), Jens Krüger, Jürgen Müller, Stephan Müller, David Schwalb, Christian Schwarz, Ralf Teusner, Christian Tinnefeld, Matthias Uflacker (now with SAP), and Johannes Wust. Working with all you people has been a tremendously rewarding experience.

I am deeply grateful to Dean Jacobs, who helped me to frame this dissertation project and has spent endless hours discussing ideas, plans and results, all out of personal interest and drive. His dedication to my particular thesis work over many years gave me the opportunity to learn a great deal from him. After all, he was probably the brightest person I have ever met.

I am thankful to SAP for the fruitful collaboration, in particular for their efforts in providing me with real-world cluster log data for experimentation. It has been an extraordinary pleasure to work with Tim Januschowski from SAP's Innovation Center, who introduced me to the art of mathematical optimization. Special thanks to Dr. Vishal Sikka for his ongoing support and guidance over the years.

Thanks to the folks at the UC Berkeley AMP Lab, where I spent a very productive time, for their hospitality and inspiring collaboration.

I would also like to thank the reviewers, Prof. Wolfgang Lehner and Prof. Alfons Kemper, for their time and effort.

A special thank you to Julia Pfaff for managing all aspects of my personal life during the last six months before submitting this thesis. Finally, I thank my parents, Cornelia and Wilfried Schaffner, for their support throughout the whole journey.

Contents

1 **Introduction** 1
   1.1 Contributions 5
   1.2 Structure 6

2 **Background and Motivation** 9
   2.1 In-Memory Column Databases 9
   2.2 Mixed Workloads 11
   2.3 Multi Tenancy 13
   2.4 Enterprise SaaS Log Data Analysis 14
   2.5 Rock: An Elastic Cluster Infrastructure for Multi-tenant Databases 16

3 **A Model for Load Management and Response Time Prediction** 19
   3.1 Benchmark Design 20
   3.2 An Empirical Model for Response Time Prediction 22
      3.2.1 Resource Consumption of Multiple Homogeneous Tenants 24
      3.2.2 Resource Consumption of Multiple Heterogeneous Tenants 27
      3.2.3 The Effect of Resource Consumption on Response Times ... 28
   3.3 Extending the Model with Batch Writes 31
   3.4 Extending the Model with Migrations 33
      3.4.1 Resource Consumption of Migrations 33
      3.4.2 Duration of Migration 36
   3.5 The Impact of Virtualization 37
   3.6 Remarks 38

4 **The Robust Tenant Placement and Migration Problem** 41
   4.1 Static Placement 42
      4.1.1 Interleaving Tenant Replicas 44
      4.1.2 Choosing the Number of Replicas 49
      4.1.3 Recoverable and Flexible Placements 51

	4.2	Incremental Placement	52
	4.3	Complexity Analysis	55
	4.4	Remarks	58

5 Algorithms for RTP ... 59
5.1 Algorithms for Static RTP ... 59
- 5.1.1 Greedy Heuristics ... 59
- 5.1.2 Meta-heuristics Based on Randomness ... 61
- 5.1.3 Exact Algorithm ... 62

5.2 Algorithms for Incremental RTP ... 63
- 5.2.1 A Framework for Incremental RTP ... 63
- 5.2.2 Greedy Heuristics ... 65
- 5.2.3 Meta-heuristics Based on Randomness ... 66
- 5.2.4 Portfolio Approach ... 66
- 5.2.5 Exact Algorithm ... 67

6 Experimental Evaluation ... 69
6.1 Tenant Trace Data Used for Experimentation ... 69
- 6.1.1 Bootstrapping Process ... 70
- 6.1.2 Tenant Sign-Ups ... 72
- 6.1.3 Choosing the Bootstrap Window Size ... 72

6.2 Evaluation of Algorithms for RTP ... 73
- 6.2.1 Comparison of Heuristics for RTP ... 74
- 6.2.2 Advanced Experiments with Robustfit-Incremental ... 82
- 6.2.3 Generic Over-Provisioning Strategies ... 88

7 Related Work ... 95
7.1 Workload Management ... 95
- 7.1.1 Workload Characterization in Database Systems ... 96
- 7.1.2 Workload Characterization in Other Areas ... 100

7.2 Data Placement ... 101
- 7.2.1 Placement Strategies in Parallel Databases ... 103
- 7.2.2 Recent Approaches for Database Replica Placement ... 108
- 7.2.3 Data Placement in Other Areas ... 110

7.3 Other Aspects of Multi-tenant Databases ... 112

8 Conclusions and Perspectives ... 113
8.1 Summary ... 113
- 8.1.1 Modeling Scan Capacity in a Column Store ... 114
- 8.1.2 Predicting 99-th Percentile Latencies ... 114
- 8.1.3 Prediction in the Presence of Perturbing Factors ... 114
- 8.1.4 Robust Tenant Placement and Migration ... 115
- 8.1.5 Algorithm Design ... 115
- 8.1.6 Algorithm Evaluation ... 116

8.2 Guidelines ... 117
8.3 Future Work ... 119

Bibliography ... 121

List of Figures

Fig. 1.1	Aggregate request rate across five calendar weeks, including Christmas	4
Fig. 2.1	NSM vs. DSM	10
Fig. 2.2	Query execution in row- vs. column-oriented databases	10
Fig. 2.3	Distribution of query types	12
Fig. 2.4	Different usage patterns for regular tenants and trial customers	15
Fig. 2.5	Database sizes for the tenants in the trace	16
Fig. 2.6	The Rock cluster architecture	17
Fig. 3.1	Schema of the Star Schema Benchmark	21
Fig. 3.2	Maximum MB scanned as a function of tenant size	24
Fig. 3.3	Maximum number of requests as a function of tenant size	25
Fig. 3.4	Scan capacity consumption for multiple tenant configurations	29
Fig. 3.5	Q-Q plot for estimated and measured 99-th percentile values	30
Fig. 3.6	Increasing response times due to periodic writes	31
Fig. 3.7	Capacity of a single instance with batch writes	32
Fig. 3.8	99-th percentile values with ongoing migrations	34
Fig. 3.9	Prediction of 99-th percentile values during migrations	36
Fig. 3.10	Duration of migration impact by size of migrated tenant	36
Fig. 3.11	Overhead of execution in a virtualized environment	38
Fig. 4.1	Comparison of placement strategies	45
Fig. 4.2	Mirrored vs. interleaved placement in Example 4.2	46
Fig. 4.3	$\mathcal{L}$ during failures on the worst server	48
Fig. 4.4	Required number of servers dependent on $r(t)$	51
Fig. 6.1	The bootstrapping process for generating realistic tenant load traces	71
Fig. 6.2	Choosing the optimal bootstrap window size W	73
Fig. 6.3	Number of active servers on an average day for selected algorithms	76
Fig. 6.4	Percentage of overloaded servers	78

Fig. 6.5	Average excess load across servers	79
Fig. 6.6	Excess load on worst server	80
Fig. 6.7	Number of invalid placements with varying migration budget	81
Fig. 6.8	Number of servers required by algorithms for static RTP	82
Fig. 6.9	MIP results for a small data set	84
Fig. 6.10	Number of active servers with a varying replication factor	86
Fig. 6.11	Temporarily overloaded servers and MB migrated with varying offset	87
Fig. 6.12	Over-provisioning strategies for avoiding overloaded servers	89
Fig. 6.13	Impact of multiple simultaneous failures	91
Fig. 7.1	Positioning of this dissertation among related work	96

List of Tables

Table 3.1	Coefficients for Eq. (3.1)	26
Table 3.2	Goodness of fit for Eq. (3.1)	26
Table 3.3	Coefficients for Eq. (3.6)	29
Table 3.4	Goodness of fit for Eq. (3.6)	30
Table 3.5	Coefficients for Eq. (3.6) with writes	32
Table 3.6	Goodness of fit for Eq. (3.6) with writes	33
Table 3.7	Coefficients for Eq. (3.6) with migrations	35
Table 3.8	Goodness of fit for Eq. (3.6) with migrations	35
Table 4.1	Maximum number of concurrent users before SLO violation occurs	47
Table 6.1	Server cost and running time of heuristics for RTP	75
Table 6.2	Improvements obtained by cplex-static and cplex-inc	85
Table 6.3	Daily server cost with varying offset	86
Table 6.4	Maximum number of unavailable tenants	94
Table 7.1	Positioning of this dissertation among parallel and Web databases	102
Table 7.2	Interleaved declustering with four nodes	105
Table 7.3	Chained declustering with four nodes	106

Chapter 1
Introduction

More and more enterprise applications are moving away from the traditional, on-premise deployment model to a hosted, on-demand model [121]. An important case here is Enterprise Software-as-a-Service (SaaS), where a service provider develops an enterprise application and operates the system to host it for many businesses. Using a SaaS solution relieves the customer of the burden of provisioning the hardware as well as operating and maintaining the application, including database configuration and tuning. On the one hand, SaaS is especially attractive for small- to mid-size businesses [41], which often lack specialized skills to efficiently perform these tasks. The service provider, on the other hand, can leverage economies of scale by automating common maintenance tasks as well as by consolidating tenants onto the same machine to improve utilization and thereby decrease costs. Such consolidation is referred to as multi tenancy [11]. Multi tenancy can be applied on multiple levels of the software stack, the database layer being one of them. As an example, already in October 2007, the SaaS CRM vendor RightNow had 3,000 tenants distributed across 200 MySQL database instances with 1–100 tenants per instance [87]. Consolidation is key in SaaS, especially considering that profit margins are commonly lower than for traditional, on-premise software.

While the economical benefit of consolidation is easily noticed when a SaaS provider rents its server resources using a public cloud infrastructure (such as Amazon EC2 [6]), one might ask whether there is an economical benefit to server consolidation in privately owned data centers. According to a study by James Hamilton [56], the constituents of the monthly cost for operating a 50,000 machine data center are (i) server amortization (54 %), (ii) power distribution and cooling (21 %), (iii) energy consumption (13 %), and (iv) general networking and infrastructure expenditures (12 %).[1] Multi tenancy leads to a reduction of the overall required number of servers, which reduces monthly amortization costs in the first

---

[1] At a scale of 50,000 servers, the expenditures for administrator salaries are negligible in comparison to these factors.

place. Moreover, given that a third of the monthly expenditures is accrued by costs related to energy consumption, powering down nodes that are momentarily not required can positively affect a SaaS provider's bottom line.

The concept of server virtualization has been successfully applied to implement such consolidation for various types of systems. The focus of this dissertation is on consolidation techniques that are specific to database systems. While multiple tenant databases could be consolidated by giving every tenant its private virtual machine (and, of course, running several virtual machines per physical server), a higher amount of consolidation can be achieved when implementing multi tenancy directly inside the database [32, 60]. The reasons for this include the overhead of running a separate operating system and database process for each tenant on a physical machine. Also, most database systems tend to allocate all available resources on a machine, which makes it hard for administrators to determine the correct virtual machines sizing. In this situation, administrators are likely to choose too conservative a sizing, which in turn results in unused potential for consolidation.

SaaS providers face a challenging trade-off between minimizing their operational cost (through consolidation) and performance as perceived by customers, especially given that on-demand enterprise applications are often required to operate within stringent performance bounds. Only so much consolidation can occur without significantly reducing responsiveness. For managing this trade-off, the service provider must address two complementary challenges:

1. *Workload Modeling:* The estimation of (shared) resource consumption in the presence of multi tenancy on a single server. This is done by characterizing the dominating resources (e.g. CPU, RAM, or disk I/O) and quantifying how much each tenant utilizes them. In order to make performance guarantees, the utilization of the dominant resources must be mapped to response times.
2. *Data Placement:* The assignment of tenants to servers in a way that minimizes the number of required servers (and thus cost). This step also entails replication of tenants for performance and high-availability.

Both challenges integrate naturally, a solution to the former being a prerequisite for the latter. The *thesis* defended in this dissertation is: "given appropriate techniques for workload modeling and data placement, a SaaS provider can operate its service in a cost-effective manner, while providing response time guarantees, even in the presence of failures."

The first challenge, the estimation of the combined resource consumption of multiple tenants on a machine, provides an indicator for the "fill level" of a server. The goal is to know in advance whether adding another tenant to a server will result in violating service-level objectives (SLOs). This problem has recently been studied for traditional, disk-based, row-oriented database systems with a focus on update-intensive workloads [32,37,76]. In this dissertation, the focus is on enterprise applications with a so-called mixed workload, which are known to have read-mostly characteristics [73]. Given that these workloads are well suited for in-memory column databases [93, 94], we develop a model for characterizing the load on an in-memory column database containing multiple tenants with different database sizes and request rates.[2] Previous cost models for in-memory databases are focused on characterizing the costs for individual queries [78, 81], while our model aims to characterize the ability of the database to handle an on-going stream of queries within pre-defined response time goals. More specifically, our model predicts how much load a server can sustain before query response times in the 99-th percentile exceed 1 s (which is our SLO). We extend this model to capture drops in capacity incurred when tenants are migrated between servers and present a method for live tenant migration in the cluster to ensure that SLOs are met.

Our approach to modeling database performance is empirical. Rather than characterizing all constituents of the system and their interactions, we extract the model from observations of a running system. As we will show, the resulting model is very accurate: our prediction of the response time in the 99-th percentile is always within 10 % of the actual value. This is a very low error, even in comparison to sophisticated statistical machine-learning techniques [47]. The reason we are able to build such a precise model using relatively simple tools is that in-memory column databases behave very predictably. We will show that, for analytical and mixed workloads, response times are proportional to the amount of data scanned by the database in a given time interval. The ability to accurately predict response times makes it possible to run servers at a higher utilization level, thereby decreasing cost.

There is considerable potential for consolidation through multi-tenancy even on a single in-memory column database instance. As an example, we analyzed a cube in the data warehouse of a large SAP customer, a Fortune 500 retail company. This cube contained all the sales records for 3 years and the fact table had approximately 360 million records. In the dictionary compressed database, this cube consumed only slightly more than 2 GB of memory, which is low given the DRAM capacities found in modern server hardware. Moreover, SaaS often targets small to mid-sized businesses with orders of magnitude smaller data sets. Our techniques are targeted for consolidating hundreds of tenant databases onto a single machine.

Given a handle on the "fill levels" of servers, the second of the above challenges can be addressed: the problem of assigning tenants to servers. This second challenge has received much less attention than the first one and, in contrast to the first challenge, considers the cluster as a whole. In existing solutions, tenant placement is performed in a fairly static manner and changes in tenant behavior over the day are not leveraged [32, 57, 58, 76, 82, 125]. If replication is considered, it is typically done using a highly structured approach such as mirroring or chained declustering. In this dissertation, we argue for a more dynamic layout of replicas across the cluster.

We present and formalize the Robust Tenant Placement and Migration Problem (RTP) and make the case for *incremental* tenant placement, driven by diurnal variations in user load. Individual tenant replicas are migrated while the tenant remains

---

[2]The present dissertation is a single-author work. Here, the use of the word "we" is an attempt to avoid recurring passive voice throughout the document. The advantage of "we" over "I," "one," or "the author" is that it (sometimes) includes the reader, which puts emphasis on the content rather than the author. As a result, hopefully, the dissertation will be more accessible to the reader.

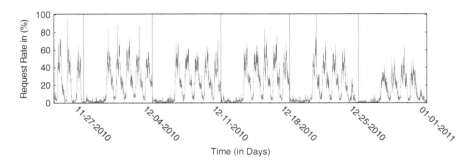

Fig. 1.1 Aggregate request rate for all tenants of hosted, multi-tenant enterprise application shown across five calendar weeks, including Christmas

on-line [38, 99]. This model allows us to make frequent, incremental changes to the tenant placement in the cluster, with the goal of running with the minimal number of servers at each point in time. This optimization is complementary to the workload management approach, which is targeted at consolidating as many tenants as possible onto a single server. In fact, incremental tenant placement is the second lever for increasing consolidation and thereby decreasing operational cost.

The potential cost savings from incremental placement are immense. We have analyzed the variations in tenant load in one of SAP's multi-tenant on-demand applications. Figure 1.1 shows a normalized view of the aggregate number of requests across all tenants using this application.[3] The chart covers a 5 week period. We can clearly identify business days and weekends in the trace. We observe that the tenants' usage of the application follows a strong 9–5 workday pattern: the number of active users per tenant and, correspondingly, request rates are elevated between 9:00 a.m. and 5:00 p.m., with a drop around lunch time. From looking at Fig. 1.1 it becomes immediately clear that provisioning for the peak load results in under-utilizing the cluster by a vast amount for most of the time.

In this dissertation, we present several novel algorithms for incrementally assigning tenants to servers in a way that servers are not overloaded (i.e. response times do not exceed a certain 99-th percentile latency; we use our workload model for this purpose) and tenants are replicated at least once. Also, tenant placement is done such that an unexpected server outage in the cluster does not result in any of the remaining servers becoming overloaded. Further, our algorithms take migration cost into account and ensure that response time guarantees are not violated as a consequence of tenant migrations.

At the core of our approach is the concept of *interleaving* replicas across nodes, which has been studied in the context of fault-tolerance for parallel databases [57]. We are also interested in tolerating server failures, but our problem is different: we

---

[3] For confidentiality reasons, we are not allowed to publish absolute requests rates. Thus, the figure only shows relative values.

are trying to minimize the number of servers required at each point in time, whereas the existing work on parallel databases assumes a fixed cluster size [57, 58, 82]. Also, in our scenario, a typical SaaS tenant is small and hence there is no benefit to horizontal partitioning, which is a prerequisite of existing interleaving strategies for parallel database systems [44, 82, 120].

Our experiments show that, with our incremental placement algorithms, operational cost for an average business day can be decreased by a factor of 10 (measured in Amazon EC2 server hours), in comparison to static placements provisioned for peak load, which are state-of-the-art. More drastic savings can be realized during longer periods of low activity, such as weekends or holidays.

Unfortunately, consolidating tenants too aggressively increases the proneness of servers to becoming overloaded due to sudden increases in tenant load. In such situations servers become temporarily overloaded and remain in this state until the load changes have been observed by the monitoring infrastructure, the tenant placement algorithm has been invoked to compute a new incremental placement, and the migrations for implementing the placement have been physically carried out. To this end, we design and evaluate several *over-provisioning strategies*. These ensure that the cost savings of incremental placement are retained for the most part, while reducing the impact of temporary overloads in terms of 99-th percentile latency to a negligible level.

When used in conjunction, the methods we propose in this dissertation enable SaaS providers to maximize consolidation in the database layer while providing performance guarantees as well as resilience towards load spikes and server failures. Therefore, our methods help to operate a cloud service efficiently.

1.1 Contributions

In summary, this dissertation makes the following contributions:

- We present a workload model for estimating the combined resource utilization of multiple tenants on a server. Our model takes the sizes and request rates of all tenants that are to be assigned to a server as input and outputs a prediction for the 99-th percentile response times that this server produces, given that all tenants run the same schema and queries. The accuracy of our predictions is greater than 90 % when response times are below 1 s.
- We extend our workload model to capture additional resource utilization incurred by batch writes and migrations. The high accuracy of our predictions remains intact, even in the presence of these processes.
- We present and formalize the Robust Tenant Placement and Migration Problem (RTP), an optimization problem aiming to minimize the number of servers required for operating all tenants over time. We show that existing bin-packing algorithms are not applicable to RTP, which makes it a new problem, and prove its $\mathcal{NP}$-completeness.

- We develop a suite of algorithms for solving RTP. We develop mixed integer programs that aim to compute exact solutions for both the static and the incremental variant of RTP. We propose several greedy heuristics, metaheuristics relying on randomness, and hybrid algorithms that combine individual strengths of multiple algorithms.
- For the purpose of evaluating our algorithms in a realistic setting, we analyze the log data of a production multi-tenant, on-demand application by SAP. We present a bootstrapping technique that, based on the variations in tenant load observed in such data, generates (large) corpuses of load traces with realistic characteristics.
- We conduct a thorough experimental study for evaluating the performance of our algorithms in terms of cost, computation time, and robustness towards unexpected load spikes. We also study the performance of our algorithms when the length of the reorganization interval between two placements decreases.
- We design and evaluate several strategies for increasing the robustness of placements towards sudden increases in load. These strategies complement our placement algorithms.
- We show that there are circumstances where increasing the number of replicas of individual tenants decreases the total number of servers. We experimentally study the question of how many replicas each tenant should ideally have.
- We study the resilience of our placements towards single and multiple simultaneous server failures, both in terms of overloaded servers as a consequence of load redistribution and tenant availability.

1.2 Structure

This dissertation is structured as follows: In Chap. 2, we give a brief overview of the technologies on top of which the workload modeling and data placement techniques presented in this dissertation are built. These include in-memory column databases, mixed workload applications, and multi-tenant architectures. Chapter 2 also motivates the need for workload modeling techniques and incremental data placement with an analysis of log data from a hosted, multi-tenant application by SAP. Finally, Chap. 2 discusses how the backend of a hosted in-memory database service with replication, multi tenancy, and an elastic number of server nodes that is dynamically provisioned, could be implemented from a cluster architecture point of view.

In Chap. 3, we address the first of the two challenges presented above. We develop a model for estimating the combined load on a server when consolidating multiple tenants onto it. This is the first main contribution of the dissertation. Since our model is of empirical rather than analytical nature, it depends on the schema and workload used for conducting experiments. We thus begin with presenting SSB-MT, our benchmark for multi-tenant SaaS applications. Afterwards, we present our workload model in Sect. 3.2. We proceed with extending the model for drops in server capacity incurred by batch writes and migrations. The chapter also discusses

1.2 Structure

the impact of virtualization on scan-intensive database workloads to underline that our model can also be applied in virtualized environments. The chapter closes with a discussion of the limitations of our workload modeling approach.

Chapter 4 addresses the second challenge. It presents and formalizes RTP, which is the second main contribution of this dissertation. We begin with discussing the static variant of RTP and present several examples of the advantages of interleaved replica placement over mirroring. Afterwards, we discuss how to obtain the minimum number of replicas per tenant. We then extend static RTP and introduce the incremental variant of the problem. The chapter closes with an $\mathcal{NP}$-completeness proof for both static and incremental RTP.

In Chap. 5, we present several algorithms for solving RPT, extending the second main contribution of the dissertation. Following the structure of Chap. 4, we begin with presenting our algorithms for static placement, which form the foundation for our incremental algorithms. Chapter 5 also introduces a framework for structuring how solutions of incremental RTP are computed, which is independent of the placement algorithms.

In Chap. 6, we evaluate our algorithms for RTP. Since our evaluation is based on the production load traces presented in Chap. 2, we begin with describing our methodology for bootstrapping these traces to generate a sufficiently large number of new traces for testing our algorithms at scale, while ensuring that the resulting set of tenants is realistic. Afterwards, we present a detailed series of experiments characterizing the performance of our various algorithms in terms of solution quality and computation time. Based on the experimental evaluation, Chap. 6 also introduces three generic over-provisioning strategies to minimize the impact of temporarily overloaded servers in response to sudden load spikes. We then show that these strategies can also be used to provide resilience towards multiple simultaneous server failures.

Chapter 7 surveys related work. Following the structure of the dissertation, we begin with discussing related work in the area of workload management and proceed with related work on data placement, afterwards. Most importantly, we contrast our work to data placement and availability techniques in the field of parallel database systems. While our focus is mainly on related work in the database community, we also discuss important related approaches from other communities.

Chapter 8 concludes the dissertation, summarizing our main findings and discussing perspectives for future work.

Chapter 2
Background and Motivation

The purpose of this chapter is to give a brief overview of the technologies on top of which the workload modeling and data placement techniques presented in this dissertation are built. We begin with an overview of in-memory column databases. We then discuss mixed workload applications, a scenario where this kind of database system is used. Afterwards, we provide background on multi-tenant architectures. We then motivate the need for workload modeling techniques and incremental data placement with an analysis of log data from a hosted, multi-tenant application by SAP. Finally, we propose an architecture for implementing the backend of a hosted in-memory database service with replication, multi tenancy, and an elastic number of server nodes that is dynamically provisioned. Our implementation of this architecture, called *Rock*, is equally applicable in public and private cloud settings. Note that this chapter does not intend to list the assumptions for our workload modeling and data placement techniques, or to frame a stringent system model; assumptions made in developing the main contributions of this dissertation are discussed in the respective chapters.

2.1 In-Memory Column Databases

Database systems are traditionally built on an n-ary storage model (NSM), in which all attributes of a tuple are physically stored together. In 1985, Copeland and Khoshafian introduced the decomposed storage model (DSM) [28]. In DSM, columns are stored physically separate from each other. The logical integrity of individual columns is preserved by the introduction of so-called surrogate identifiers, denoted as "sID" in Fig. 2.1. Surrogate identifiers are either stored explicitly, which leads to additional storage consumption, or they can be implicit, i.e. they can be inferred from the positions of the values in a column.

DSM has been implemented in a multitude of projects, for example, MonetDB [18], C-Store [107], or BigTable [23]. Columnar storage has also been

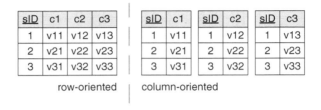

Fig. 2.1 NSM vs. DSM

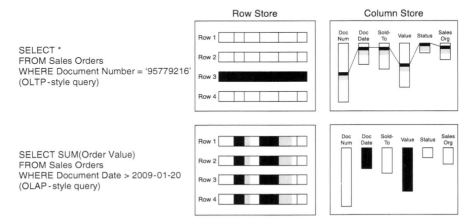

Fig. 2.2 Query execution in row- vs. column-oriented databases (From Plattner [94])

implemented in a range of commercial products such as Sybase IQ [108], Vectorwise [129, 130], or SAP HANA [42, 61, 93, 98, 103].

Figure 2.2 illustrates the differences in query processing for row- and column-oriented databases. When executing point queries with full projectivity, such as in the upper SQL statement in Fig. 2.2, row-oriented databases can simply look up the requested tuple (typically using an index) and read the values of all attributes of the tuple en bloc. A column-oriented database must perform a so-called "positional join" for each attribute to re-assemble the complete tuple. When executing a "set-oriented" query (i.e. a query that is attribute-focused rather than entity-focused [1]), only a small number of columns in a table might be of interest for a particular query, as is the case for the lower SQL statement in Fig. 2.2. Here, the column store accesses only the columns that must be read for processing the query, while the rest of the table can be ignored. All values of a column can be scanned en bloc. This has three major advantages: (i) only few data that is not relevant for computing the query result is read; (ii) pre-fetching mechanisms can exploit the spatial locality of the values, which reduces the number of cache misses and ensures that the CPU does not stall while waiting for data; and (iii) as a consequence of the data in the columns being compressed, more values fit on a single cache line. This is in contrast to the row-oriented model, where all tuples matching the filter condition are inspected one after another, each time extracting the required attribute values from the tuple. This has one major disadvantage: the cache line holding the value of a required attribute is likely to also hold attribute values of irrelevant tuples. Consequently, large amounts of data not required for answering the query are read, leading to a polluted cache. The areas that are shaded in light grey in Fig. 2.2 illustrate data, which is loaded into the caches but is not useful for query processing. Abadi et al. demonstrate that the advantages of a column store database architecture cannot be achieved when emulating a column store in a row-oriented database, i.e. by using separate tables and indexes for each column of a logical table [2].

Many columnar databases store values in a compressed format, as we illustrate in Fig. 2.2 using different column lengths. This is done mainly for two reasons: (i) saving space and (ii) increasing performance. Compressed columns obviously consume less space on disk or in-memory. Decompressing data upon read causes additional load on the CPU, however. A trade-off that balances the compression ratio and the cost for de-compression must be found. While CPU capacities grow at a rate of 60 % per year (measured in the number of transistors, not clock rate), the access latency of main memory decreases by less than 10 % per year [80]. This trend favors the usage of compression techniques requiring more effort for de-compression. Compression techniques exploit redundancy within data and knowledge about the data domain for optimal results. Compression applies particularly well to columnar storage, since all data within a column (i) has the same data type and (ii) typically has similar semantics and thus a low information entropy (i.e. there are few distinct values in many cases). For attributes with a high number of distinct values, dictionary encoding can significantly reduce the amount of space required for storing an attribute. The basic idea is that frequently appearing values are replaced by smaller symbols. Depending on the size of the dictionary for an attribute, the lowest number of bits necessary for representing all entries in the dictionary can be used for storing the keys of the dictionary.

SAP HANA, the database used for the experiments in Chap. 3 of this dissertation, mainly uses dictionary compression for individual columns and bit-vector encoding for storing the keys into the dictionaries.

2.2 Mixed Workloads

Traditionally, enterprise applications have been conceptually and physically split into systems for on-line transactional processing (OLTP) and on-line analytical processing (OLAP). On the database level, OLTP applications are widely believed to be characterized by a balanced number of read and write queries, where most of the read queries only touch single or few rows. OLAP systems, in contrast, are believed to consist mainly of read queries that use few attributes but a large number of rows, as is the case for typical aggregation queries.

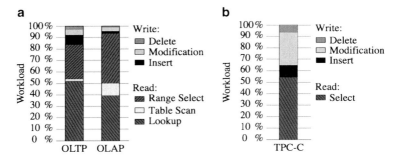

Fig. 2.3 Distribution of query types (From Krüger et al. [73]). (**a**) Analyzed customer systems. (**b**) Standard benchmark

Krüger et al. have analyzed 12 SAP ERP Business Suite customers with 74,000 tables each [73]. They have studied what types of queries these systems issue against the database, for both the customers' OLTP and OLAP systems. The results are shown in Fig. 2.3a. The results of the analysis of the OLAP systems are in line with the characterization of OLAP given in the previous paragraph: write queries account for less than 10 % of the total workload. Most of the read queries are scan-intensive, i.e. range selects or explicit table scans. The workload of the analyzed OLTP systems differs from the characterization of OLTP given above: write queries account for only 17 % of the total workload, most of which are inserts. More than 50 % of the queries are lookup queries (similar to TPC-C [112], shown in Fig. 2.3b) and more than 30 % of the queries are range selects and table scans. TPC-C, the standard benchmark for OLTP systems, does not contain range selects and table scans at all, as shown in Fig. 2.3b.

The enterprise software on which the analysis by Krüger et al. is based on was first released almost two decades ago. In more recent enterprise systems, there is a trend towards providing more ad-hoc operational reporting capabilities directly on the transactional data [93, 98]. Also, there are applications which cannot be clearly assigned to either the OLTP or the OLAP category, such as "available-to-promise" or planning applications.[1] These applications frequently insert data into the database, but also perform complex aggregation queries as found in OLAP systems. Such applications exert a *mixed workload* of transactional and analytical queries on the database [17, 26, 72]. Therefore, it is reasonable to expect that the fraction of range queries and table scans in OLTP workloads increases in the future.

---

[1] Available-to-promise (ATP) is a process for determining whether a given quantity of a product can be shipped to a customer by a desired delivery date. ATP occurs during order-entry transactions and during shipment rescheduling, where certain customers or orders are prioritized in case of stock shortages.

2.3 Multi Tenancy

Multi tenancy encompasses methods for consolidating multiple customers (i.e. tenants) of a hosted application into the same system with the goal of reducing operational cost. The term "multi tenancy" has been coined by the hosted application service provider salesforce.com who implements multi tenancy in multiple layers of their software stack [121]. Multi tenancy can also be utilized in the database layer such that a single database is shared by multiple tenants. According to Jacobs and Aulbach, multi tenancy in the database layer can be realized by adopting either a shared-machine, shared-process, or shared-table approach [60], which we summarize briefly in the following.

In the shared-machine approach, each tenant runs in a separate database process (potentially inside its own virtual machine). Isolation among tenants is very strong in this approach. The downsides are (i) that resource management (i.e. the scheduling of queries or memory management) is not done holistically for the whole server; (ii) the application server must maintain an individual socket connection for each database process (although all processes reside on the same server); and (iii) cluster management operations cannot be applied in bulk across all tenants.

In the shared-process approach, all tenants share a single database process, although each tenant has their private instance of the database schema. This can be realized using tablespaces or a special naming scheme for the database tables. The shared process approach does not have the downsides of the shared-machine approach but isolation among tenants is weaker. According to Aulbach et al. [11], some database systems, such as IBM DB2, allocate a fixed amount of buffer pool memory per table, so that maintaining a large number of tables can become problematic.

In the shared-table approach, all tenants share one database process and one instance of the schema. To do so, a `tenant_id` column is added to each table. In this variant, the overhead for a new tenant is minimal and access can typically be made efficient by using an index for the `tenant_id` column. Isolation is weakest in the shared-table approach. In fact, the shared table approach may be problematic from a security point of view, since typically access rights in databases can only be granted on the level of tables and not on individual rows. Oftentimes, tenants require that local modifications to the database schema can be made. Such schema extensibility is harder using the shared-table approach than for the others, since it requires using generic columns (e.g. `varchar1`) and the application must provide a mapping between application fields and generic columns in the database. Further, the shared-table approach instigates resource contention. One example for such contention is the case where a query of a tenant results in a full table scan. Then, the data of all tenants must be read. Another example is that executing administrative tasks for a single tenant (e.g. migration) negatively affects the performance of all other tenants.

2.4 Enterprise SaaS Log Data Analysis

In order to better understand the usage patterns of Enterprise SaaS applications, we analyzed real world load traces obtained from a production multi-tenant, on-demand application by SAP. These traces are the anonymized application server logs of 87 randomly selected tenants in Europe over a 4 months period. We also have additional statistics such as the tenants' database size. Figure 1.1 (on page 4) shows a normalized view of the aggregate number of requests across all tenants over a 5 week period. As already stated in Chap. 1, we can clearly identify workdays and weekends in the trace and observe strong 9–5 workday pattern. Request rates are elevated between 9:00 a.m. and 5:00 p.m., with a drop around lunch time.

Our analysis of the traces also reveals that holidays, such as Christmas, have the expected effect of lowered request rates for a limited period of time: the last week shown in Fig. 1.1 is the week between Christmas and New Year's Day. Tenant activity in this week still conforms to the same periodic pattern as in the weeks before but exhibits a much lower amplitude. Also, as Christmas Eve in 2010 fell on a Friday, it comes as no surprise that the load on this particular Friday is considerably lower than the other Fridays.

Most individual tenants exhibit a load pattern similar to the aggregate load across all tenants, shown in Fig. 1.1. However, we found a number of interesting particularities in analyzing the load traces of the individual tenants. Figure 2.4 shows the load traces of four different individual tenants.[2] While most tenants exhibit a load pattern as shown in Fig. 2.4a, there are also tenants in the trace which only use the system occasionally (Fig. 2.4b). With incremental tenant placement as proposed in this dissertation, fewer resources can be allocated to those tenants during times of inactivity, thereby temporarily reducing server cost. A non-negligible number of tenants emerge suddenly, are active for 12 weeks, and then abruptly become inactive (Fig. 2.4c). During this time they behave similar to regular clients. Other short-lived tenants use the system actively for 2–3 weeks, then become inactive for a considerable amount of time (e.g. for 2 weeks), and suddenly resume activity for 6 weeks periods (Fig. 2.4d). After investigating these tenants more carefully, we discovered that these were mainly demo and training systems. Capacity planning often neglects those systems. The behavior of tenants in trial periods is particularly hard to predict, which, in part, motivates the incremental placement algorithms presented in this dissertation.

Another interesting insight drawn from the sample is the distribution of the tenants' database sizes, shown in Fig. 2.5. Tenant sizes follow a long-tail distribution rather than a self-similar distribution as is often assumed [54]. Hence, a few tenants are significantly larger than the rest, but the vast majority of tenants have approximately the same size. The larger tenants also have significantly more active users and account for the majority of the load in the cluster.

---

[2] Again, the figure only shows relative values.

2.4 Enterprise SaaS Log Data Analysis

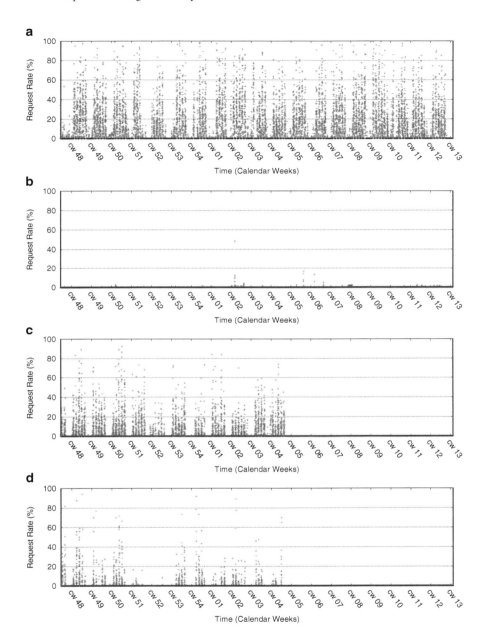

Fig. 2.4 Different usage patterns for regular tenants and trial customers. (**a**) Regular tenant. (**b**) Tenant with low activity. (**c**) Demo system used actively until end of trial. (**d**) Demo system used only at beginning and end of trial

Fig. 2.5 Database sizes for the tenants in the trace

These results motivate the need for continuous response time prediction in the presence of multi tenancy as well as incremental data placement. Furthermore, we will use these traces to design a realistic testbed for evaluating the different placement techniques proposed in this dissertation (see Chap. 6).

2.5 Rock: An Elastic Cluster Infrastructure for Multi-tenant Databases

Having motivated the need for an elastic cluster architecture, we describe the Rock clustering infrastructure in this section, which has been implemented in the course of this dissertation project. It provides a backend infrastructure for running hosted database services on a collection shared-nothing servers (*processing nodes*). These servers run instances of the SAP HANA [42, 103] database. Rock manages multi-tenancy, including replication of tenant data and fault tolerance, and tries to continuously minimize the number of active servers. Rock can be used in conjunction with a public cloud service, such as Amazon EC2 [6], or in a private cloud setting, e.g. in the data center of a hosting provider. In the latter case, third-party tools such as Eucalyptus [40] can be used to realize dynamic provisioning in a physical cluster.

Figure 2.6 illustrates the Rock architecture. Rock is intended to support analytical applications or applications with a mixed workload. In Rock, multiple tenants use the same application. All tenants have their private data, which is replicated at least once. All replicas are *active*, i.e. used for query processing. A tenant has multiple users which concurrently use the application. Read requests are submitted to the cluster by the application. There can be multiple simultaneous read requests per tenant, depending on the tenant's current number of active users. The application submits write requests in serial batches. There is only one batch writer per tenant. The three principal components of the Rock architecture are: the *cluster leader*, the *router*, and the *instance managers*. Each processing node runs one instance manager

2.5 Rock: An Elastic Cluster Infrastructure for Multi-tenant Databases

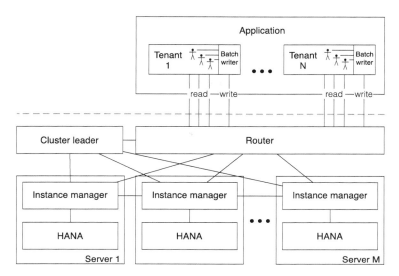

Fig. 2.6 The Rock cluster architecture

process and each instance manager is paired one-to-one with a local HANA instance to which it forwards requests. In the following, we describe these three components in more detail.

The cluster leader exists only once in the landscape and assigns tenants to servers. Each tenant replica is assigned to one instance manager and each instance manager is responsible for the data of multiple tenants. The cluster leader maintains the placement information in a *cluster map*, which it propagates to the router and the instance managers such that all components always share a consistent view of the landscape. It monitors the liveness of nodes, starts and stops servers, makes placement decisions, and triggers migrations of tenants between servers accordingly. The cluster leader is not directly involved in request processing. Since the cluster leader is a single point of failure, it must be made highly available. This could be done by running a cluster leader process on all active servers in the cluster and putting a protocol in place according to which the servers agree on a master process. Similarly, all servers in the cluster must agree on one version of the cluster map as the current version. The industry standard for addressing such problems is to use a distributed metadata store such as Chubby [19] or Zookeper [110], both of which use the Paxos algorithm [75] for keeping the metadata in sync across all sites.

The router accepts requests from outside the cluster and, based on the cluster map information, forwards them to an appropriate instance manager. It hides the physical location of a tenant's database. The router balances the load across the replicas of a tenant in a round-robin fashion. If a tenant replica becomes unavailable in consequence of a server failure, the requests are balanced across the remaining live replicas. The router also sends query results back to the application. The router collects load statistics such as requests per tenant and second. This information is attached to the query results returned to the router by the instance managers. The router assumes that write queries are serialized (in the application layer) and totally ordered on a per tenant basis, in the sense that they carry a unique, consecutive version number.

Instance managers act as an interface to the local HANA instances on the server. Their main task is to manage the distribution of write requests across the replicas of a tenant. For each tenant, the instance managers holding a replica of the tenant are organized in a logical ring, similar to Amazon Dynamo [34]. Any node in the ring can accept a write request from a router and initiate the write propagation process. When an instance manager receives a write request, it writes to the local HANA instance and, in parallel, forwards the request to its successor node in the ring. Write requests must be persistently written to i out of n nodes in a ring before the node that originally received the write request returns a confirmation to the router. The write request is then asynchronously replicated to the nodes $i + 1$ to n. Rock allows to set i and n according to the requirements of the application. A common setting for enterprise applications is $i = n = 2$, i.e. all tenants have two replicas which are strongly consistent. We assume writes come from a single source per tenant and that there are no concurrent writes. We also require that writes are sequentially numbered at the source. We require that writes are applied in the same order on all replicas. Out-of-order writes at a server are delayed until all previous writes have been received. Read consistency, which is required to support multi-query drill down into a data set, is ensured by using multi-version concurrency control (MVCC) based on snapshot isolation [13], which is natively supported in HANA. Read requests are always issued in relation to a specific version number.

As mentioned above, each instance manager is paired with a local HANA instance, which is shared by multiple tenants. Inside these HANA instances, we chose to implement multi tenancy using the shared-process approach. In our case, shared-process provides the best trade-off between overhead per tenant and the performance of tenant migrations. When tenants have their private tables, migration can be implemented on file system level. This would not be possible with a shared-table approach, where all tuples belonging to a tenant must be retrieved through the SQL layer, which incurs additional overhead.

The remainder of this dissertation focuses on the main contributions, which are in the areas of workload modeling and data placement. In Chap. 3, we will introduce a model for predicting the 99-th percentile response times produced by a server in the presence of multi tenancy. In terms of the Rock architecture, the involved components are instance managers and HANA instances. In Chap. 4, we investigate multi tenancy from the perspective of the whole cluster and present an optimization problem for replicated tenant placement. In Chap. 5, we propose multiple data placement algorithms for solving this problem. In Rock, these algorithms would be run in the cluster leader component.

Chapter 3
A Model for Load Management and Response Time Prediction

In this chapter, we develop a model for estimating the combined load on a server when consolidating multiple tenants onto it. In doing so, we consider a server to be overloaded when query response times exceed a certain threshold.

To build such a model, one could adopt either an *analytical* or *empirical* approach. Analytical models are typically based on queuing networks and require capturing all building blocks of the database architecture that are involved in query processing, including all control and data flows. Additionally, analytical models require making assumptions regarding the distributions of wait times at the various nodes in the queuing network (so-called service center wait times) as well as the inter-query arrival times. A specification of the underlying hardware and all relevant operating system aspects—such as the scheduling of parallel threads—must also be included in the model. Given the large number of assumptions involved in building analytical models, they are often unable to accurately predict response times of complex systems.

When building an empirical model, in contrast, a black box view of the database is adopted and observations about the response times are made based on experiments. The experimental results are then generalized into a model which captures all important aspects of the experiment configuration (e.g. request rates or multi-programming level) as parameters. The downside of the empirical approach is that experiments must be repeated when an aspect of the system has changed, such as the workload or the hardware. In this dissertation, we adopt an empirical approach to modeling database response times. Our model, which will be presented in the following, was first described and published in [99].

Given our empirical approach to workload modeling, we will begin with describing the workload with which we benchmarked the database system. Afterwards, we will construct our empirical model in multiple steps, eventually extending it with batch writes and migration costs. Finally, we will discuss important limitations of our model and discuss its applicability in other contexts than the used server infrastructure and benchmark.

3.1 Benchmark Design

In the following, we describe the workload that has been used in our experimental study. As discussed in Chap. 2, our focus is on in-memory column databases running enterprise applications with a mixed workload. Since dedicated benchmarks for mixed workload processing (e.g. [17, 26, 46]) have only been available since 2011, we designed our own benchmark, called SSB-MT.

SSB-MT is a modification version of the Star Schema Benchmark [91] (SSB), which is itself an adaptation of TPC-H [114]. The decision for using SSB was also influenced by the notion of *query flights* in SSB; the queries in SSB are grouped into four flights, with three to four queries each. Query flights are modeled after a user "drill-down" operation into a data warehousing data set, i.e. all queries compute the same aggregate measure but use different filter criteria on the dimensions. We believe that this is actually realistic user behavior and use the query flights as the basis for modeling concurrent users in SSB-MT as follows.

A user is a thread that executes all four query flights sequentially. A new read transaction is started at the beginning of each query flight. After receiving a response, a user waits for a fixed think time before submitting the next query. Following [104], which studies user behavior for web applications, we draw user think times at random from a negative exponential distribution with a mean value of 5 s. To prevent the "caravaning" of user threads, the first query that each user fires is offset by a random amount of time. A tenant in SSB-MT is comprised of one instance of the data model (i.e. one set of tables) and multiple users. The size of the data set and the number of users can be varied individually per tenant.

In contrast to conventional benchmarks, we do not report the end-to-end wall time for all queries in the workload as the result. Instead, we report how many concurrent users a system can support before a certain performance SLO is violated, since users are the basis of pricing and revenue in a SaaS setting.

Figure 3.1 shows the data model of SSB, which we re-use as is in SSB-MT. Since SSB is itself not very well known, we summarize the most important differences of its data model in comparison to TPC-H:

1. The TPC-H tables `lineitem` and `orders` are combined into a single table called `lineorders`. This change transforms the TPC-H data model from 3rd Normal Form (3NF) into a star schema, which is common practice for data warehousing applications.
2. The TPC-H table `partsupp` is dropped because it contains data on the granularity of a periodic snapshot, while the `lineorder` table contains data on the finest possible granularity: the individual line items. It is sufficient to store the data on the most fine-grained level available and to obtain numbers on a more coarse granularity by dynamic aggregation.

3.1 Benchmark Design

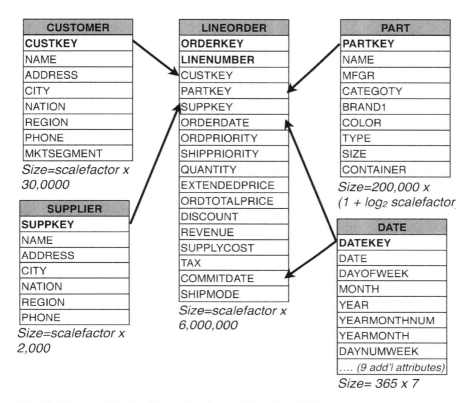

Fig. 3.1 Schema of the Star Schema Benchmark (Taken from [91])

To produce data for our experiments, we used the SSB data generator.[1] As stated in Sect. 2.5, each tenant has their own private tables; thus, there is one instance of the SSB data model per tenant. To determine realistic database sizes for the individual tenants, we analyzed the data warehouse of a large SAP customer (a Fortune 500 retail company). We found a cube (i.e. a set of database tables organized in a star schema layout) containing all sales records of this company for analysis, a similar scenario as modeled by SSB. For each calendar year, the cube contains approximately 120 million records. Since the focus of this dissertation is on SaaS, which targets small- to mid-size businesses, this number of sales records must be scaled down appropriately. Assuming that the number of sales records is proportional to revenue, we scale down the number of sales records by a factor of between 20 and 200.[2] In SSB-MT, this results in fact tables varying in size from

---

[1] http://www.cs.umb.edu/~poneil/dbgen.zip

[2] These scale factors have been chosen based on the following intuition. According to the European Commission, a small- to mid-size business has revenues of up to 50 million EUR [41],

600,000 rows (SSB scale factor 0.1) to 6,000,000 rows (SSB scale factor 1) for the different tenants. The dimension tables increase linearly in size as the fact tables grow.

TPC-H defines two *refresh functions* that change a given percentage of the rows in the fact table. They are supposed to be performed between two runs of the benchmark in order to prevent result caching. Their execution times are not measured, which means that they do not contribute to the overall pricing of the benchmark. SSB, in contrast, has no notion of writes. We use the TPC-H refresh functions to model batch writes in SSB-MT. For each tenant, we perform one batch write every 5 min. Each write increases a tenant's fact table by 0.05 % of its current size. Our decision to submit writes in batches is also motivated by practice. In SAP's enterprise systems, the application server submits data to the databases in a batch process that is decoupled from the end-user transaction on the application server. Consequently, the execution times of the writes themselves are not measured in our benchmark.

All experiments were conducted on *m1.large* instances on Amazon EC2 [6], which have two virtual compute units (i.e. CPU cores) with 7.5 GB RAM each. Given the tenant sizes described above, a server is typically packed with 14 to 27 tenants. As a point of comparison, the SaaS CRM vendor RightNow maintains between 1 and 100 tenants in a typical production MySQL database instance [87].

The results presented in this chapter are dependent on the specific workload and hardware choices described above. Nevertheless, using the methodology for obtaining an empirical model for response time prediction adopted in the following, an empirical model can be created for other workloads and hardware (as we will also discuss in Sect. 3.6).

3.2 An Empirical Model for Response Time Prediction

Typically, the determining factors that influence query performance in a database system are (i) the workload and (ii) the hardware. In the presence of multi tenancy, the number of concurrently active tenants sharing a server is a third component influencing query performance. We study the interplay of these three factors and construct a model for predicting 99-th percentile response times in the presence of multi tenancy.

In order to empirically model the resources of a server in a cluster of in-memory column databases, we need to capture the amount of work done by an individual server in a given amount of time. As we will see in the following, for SSB-MT, response times are proportional to the amount of data being "scanned" on the server

---

while revenues vary between 4.7 billion and 453 billion USD within in the Fortune 500 [20]. These ratios suggest scale factors in the range of 100–10,000. Our tenant datasets are thus—if anything—too large.

per second. Our hypothesis is that the product of table size and number of requests per second is the determining factor for performance in our case and can serve as a metric for indicating the response times produced by a server. We define the capacity limit of a server as the point at which its "scan capacity" is exhausted and, in turn, performance SLOs are violated.

The performance SLO we assume in this dissertation and which we have built into our model is defined as follows. We require that 99 % of all queries have a response time of less than 1 s over a 10 min sliding window. That means, after sorting all queries observed within the last 10 min by their response times in decreasing order, the top 1 % queries with the highest response times are removed. Among the remaining queries, the one with the highest response time is called the *99-th percentile value*. 99-th percentile response times have also been used as a performance metric by other distributed data management systems such as Amazon Dynamo [34]. Note that enforcing a certain 99-th percentile response time is a much stronger performance guarantee than asserting a specific average response time: when looking at response times as a statistical distribution, the values around the 99-th percentile can be considered outliers given the typically much smaller mean value of the distribution. Defining a limit on outlier values naturally leads to more stable and predictable response times, which in turn increases end-user confidence. For example, in our experiments, enforcing a 99-th percentile value of less than 1 s led to stable average response times in the range between 120 and 200 ms. We chose a 99-th percentile value of 1 s, since, according to Nielsen [88], 1 s is about the limit for a user to his or her attention and not lose the feeling of operating directly on the data.

We study how many tenants can be consolidated onto a server without an aggregate consumption of scan capacity beyond the point where the server fails to meet the performance SLO. In other words, we are interested in increasing the query throughput to the maximum level without violating the performance SLO. We model how much scan capacity a set of tenants of arbitrary size and with different request rates consume. The underlying observation is that processing aggregation queries of larger tenants takes longer than processing aggregation queries of smaller tenants because more data needs to be scanned. Our assumption, which is to be verified in the following, is that scan capacity consumption can be described as a function of request rate and size for each tenant.

To examine the maximum bandwidth consumption before SLO violation depending on request rate and tenant size, we conducted several tests using SSB-MT as described in the previous section. Given the analytical nature of the queries in this benchmark, the database performs full table scans or range selects for most queries. We therefore hypothesize that our results can be generalized to arbitrary scan intensive workloads, such as all analytical and mixed workload processing commonly found in enterprise applications.

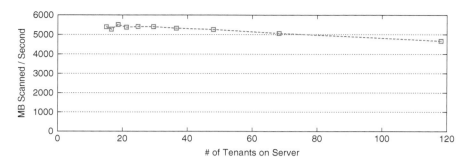

Fig. 3.2 Maximum scan capacity consumption before violating performance SLOs as a function of tenant size

3.2.1 Resource Consumption of Multiple Homogeneous Tenants

In this section, we investigate the case where all tenants on a server have the same size and request rate. We say that tenants are *homogeneous*.

In the following experiment, a single server is filled with a group of homogeneous tenants. The total number of users (and thus requests) is distributed equally among all tenants. The server is filled up such that only 20 % of its main memory is used. The resulting amount of memory is divided by the tenant size t_s. Thus, the server contains fewer or more tenants depending on the chosen value for t_s. We vary t_s from 24 to 204 MB (i.e. from 600,000 to 6,000,000 rows). The number of requests per tenant, denoted by t_r, is increased until the performance SLO is violated. For simplicity, we do not include batch writes in this initial experiment. Figure 3.2 shows the maximum throughput that the server can achieve without violating the performance SLO. Tenant size is varied along the x-axis of Fig. 3.2: small tenant sizes result in a high number of tenants on the server and vice-versa. Throughput, expressed in megabytes scanned per second, is shown on the y-axis. For calculating the amount of megabytes scanned per second, we multiply the number of requests processed by the server per second, t_r, with the tenant size, t_s. The result is an approximation of the amount of scan capacity utilized per second. We observe that the maximum number of megabytes that the server can scan before violating our performance SLO is independent of tenant size in principal. When increasing the number of tenants by a factor of 10, throughput decreases by 11 %. This slight decrease in throughput originates from a small overhead associated with processing individual requests. This overhead is mainly due to the time spent with starting and stopping read transactions, which occurs at the beginning and the end of each query flight in SSB-MT. When tenants are smaller, individual requests are processed faster, and, consequently, more transactions are started within a given period of time than for larger tenants. The main insight that can be drawn from Fig. 3.2 is that, from a

3.2 An Empirical Model for Response Time Prediction

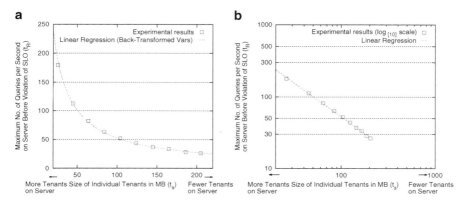

Fig. 3.3 Maximum number of requests before violation performance SLOs as a function of tenant size. (**a**) Normal scale. (**b**) Log-log scale

perspective of SLO violations, it does not matter whether the scan capacity of the server is consumed by a few large tenants or by many small tenants.

Our initial modeling goal is to predict the maximum number of requests that a server can process when tenants are homogeneous given the size t_s of the homogeneous tenants. We thus investigate the relationship between t_s and the maximum number of requests across all tenants on the server before violating the performance SLO, denoted as t_R. The capital R emphasizes that t_R is the sum of all t_r. Figure 3.3a shows this relationship. Figure 3.3a offers a different visualization of the data presented in Fig. 3.2. In Fig. 3.3a, tenant size t_s is shown on the x-axis, while t_R is shown on the y-axis. The variables t_s and t_R are inversely proportional: if t_s is divided in half, then t_R approximately doubles. At the same time, the product of t_s and t_R remains almost constant for all (t_s, t_R) pairs, as we have shown in Fig. 3.2. Thus, the graph in Fig. 3.3a is a hyperbola.

The relationship between tenant size t_s and request rate t_R, shown in Fig. 3.3a, can now be captured mathematically. This is done by fitting a function through the data points in Fig. 3.3a. Since the graph in Fig. 3.3a is a hyperbola, it can be described using the general form of a power function, shown in Eq. (3.1).

$$t_R = p_1 \cdot t_s^{p_2} \tag{3.1}$$

To achieve a higher accuracy in the curve fitting process, we use a modeling trick and transform both axes of Fig. 3.3a using the natural logarithm. In the range of t_s we are interested in (i.e. 24–204 MB), the logarithmically transformed function is no longer exponential but linear, as shown in Fig. 3.3b. To fit a curve through the data points in Fig. 3.3b, we must assign values to the parameters of the function shown in Eq. (3.2).

$$\log t_R = \log p_1 + p_2 \cdot \log t_s \tag{3.2}$$

Table 3.1 Coefficients for Eq. (3.1)

Coefficient	Value	95 % confidence bounds	
		Lower	Upper
p_1	3.9679×10^3	3.4599×10^3	4.5505×10^3
p_2	-0.9382	-0.9679	-0.9084

Table 3.2 Goodness of fit for Eq. (3.1)

Sum of squares due to error (SSE)	58.8590
R^2 (coefficient of determination)	0.9972
Adjusted $\bar{R}^2$	0.9972
Root mean square error (RMSE)	2.5573

The coefficients p_1 and p_2 can now be derived by performing a linear regression and then reverting the logarithmic transformation. Table 3.1 shows the values that have been estimated for the parameters p_1 and p_2. By fitting the linear function obtained from the log-log transformation, we were able to achieve a higher accuracy in the curve fitting process; the root mean square error was more than a factor 3 lower than for performing a non-linear curve fitting procedure on Eq. (3.1) directly.

As shown in Table 3.2, the function given in Eq. (3.1) with the coefficients from Table 3.1 is a good fit for the data points shown in Fig. 3.3a. The table shows the most important quality indicators of the regression process after inverting the log-log transformation. The adjusted $\bar{R}^2$ value (the so-called *the degree of freedom adjusted R^2 value*) is generally the best indicator of the fit quality in the presence of multiple coefficients. The closer this value is to one, the higher the quality of the fit. The root mean square error (RMSE) represents a standard error of the fit across all available data points. The RMSE has the same units of measurement than the quantity being estimated. In our case, the maximum number of requests per server without violation of the performance SLO can be estimated with an RMSE of 2.5573 queries per second.

We have shown that, for homogeneous tenants, the number of megabytes scanned per second (i.e. the consumption of scan capacity) is the determining factor for estimating whether or not a server can maintain our performance SLO of 1 s in the 99-th percentile. Based on the observation that scanning larger tables consumes more scan capacity than scanning smaller tables, we have shown that higher request rates can be achieved for small tenants before our SLO is violated. Further, the relation between tenant size and the maximum number of requests on a server before SLO violation is not linear. We have captured this relationship using a curve fitting approach, which allows us to estimate the maximum number of requests per second that a server filled with homogeneous tenants can sustain without violating performance SLOs. This result is the first component of our workload model.

3.2.2 Resource Consumption of Multiple Heterogeneous Tenants

In the following, we investigate the case where the tenants on a server have arbitrary sizes and request rates. We say that tenants are *heterogeneous*.

Our goal is to predict a server's ability to maintain the performance SLO based on tenant sizes and request rates; thus, we need to construct a function that takes the size and request rate of a single tenant as input and returns, as output, the fraction of the server's resources that this tenant will consume. The resource utilization predicted by this function shall be based on Eq. (3.1). If the sum of the values of this function for all tenants on a server is greater than a certain threshold, we predict that the server will produce SLO violations. In the following, we construct such a function $\ell : \mathbb{R} \times \mathbb{R} \to [0..1]$ from Eq. (3.1). The function ℓ takes values in the interval $[0..1]$, where a function value of 1 shall correspond to the situation where so many resources are consumed that the performance SLO is violated. A function value of zero, in contrast, shall indicate that no resources are being consumed at all. In the previous experiment, the total number of requests on a server t_R denoted the point where SLO violations begin to occur. Exactly at that request rate, our function ℓ shall thus assume a value of 1. We re-arrange Eq. (3.1) as follows:

$$t_R = p_1 \cdot t_s^{p_2}$$

$$t_R = \frac{p_1}{t_s^{-p_2}}$$

$$1 = \frac{t_s^{-p_2} \cdot t_R}{p_1} \tag{3.3}$$

While Eq. (3.1) has only one independent variable (t_s) and one dependent variable (t_R), we want to construct ℓ in a way that both tenant size and request rate are independent variables (i.e. input variables) in our function ℓ. The output of ℓ is the fraction of scan capacity consumed by an individual tenant of size t_s and request rate t_r. We define ℓ as follows:

$$\ell(t_s, t_r) := \frac{t_s^{-p_2} \cdot t_r}{p_1} \tag{3.4}$$

As a consequence of the transformation of Eq. (3.3), the value of ℓ is one for all points on the fitted curve in Fig. 3.3a; thus, $\ell = 1$ when t_R is maximal. Consequently, the domain of ℓ is restricted as follows. Any input pair (t_s, t_r) of ℓ must satisfy the constraints $0 \leq t_s \leq t_{s\max} = 205$ (the largest tenant size tested in our experiments) and $0 \leq t_r \leq t_R$.

We hypothesize that ℓ as defined above approximates the scan capacity consumption for a single tenant. We will experimentally verify this hypothesis in the following section.

We further hypothesize that ℓ is additive across multiple tenants. If this hypothesis was true, then the sum of all values of ℓ for all tenants on a server can be used for predicting whether the server can handle the aggregate load of all tenants assigned to it without violating performance SLOs. We shall also verify this hypothesis in the next section. For notational convenience, we define the total value of ℓ across all tenants on the server as follows:

$$\mathcal{L} := \sum_{t \in T} \ell(t_s, t_r) \qquad (3.5)$$

In this section, we constructed a function ℓ for calculating the fraction of a server's resources consumed by an individual tenant, using the tenant's size and request rate as input. We further assumed that ℓ is additive across all tenants on a server. In the following, we will experimentally test the hypotheses that (i) the function ℓ is an accurate predictor for the amount of resources that a single tenant consumes on a server and (ii) that ℓ is additive across all tenants on a server.

3.2.3 The Effect of Resource Consumption on Response Times

In the following experiment, we analyze four different sets of heterogeneous tenants. The sizes of the tenants in a set are chosen randomly. The total size of all tenants across the four heterogeneous sets (i.e. configurations) are 1.5, 2.0, 2.6, and 3.2 GB. While the sizes of individual tenants remained constant throughout all runs, the request rate was varied, starting from a very low request rate for all tenants and increasing beyond the point where the server violates the SLO. Again, all experiments are conducted without writes. Figure 3.4 shows the aggregate resource consumption of all tenants given the current request rate on the x-axis. The resource consumption is computed using our function $\mathcal{L}$, defined in Eq. (3.5). The y-axis of the figure shows the response times in the 99-th percentile measured for the corresponding value of $\mathcal{L}$.

We can draw two important observations from Fig. 3.4.

1. For values of $\mathcal{L}$ greater than one, denoted by a vertical line, the 99-th percentile value increases beyond 1,000 ms, denoted by a horizontal line.
2. All 99-th percentile values align on a curve, regardless which of the heterogeneous tenant sets a data point belongs to.

These observations verify our hypothesis that $\mathcal{L}$ is an accurate predictor for the amount of resources consumed by multiple tenants on a server. It also shows that $\ell(t)$ is additive across multiple tenants on a server. The fact that all data points seem to align on a curve suggests that $\mathcal{L}$ can also be used for predicting response times produced by a server in the 99-th percentile. In the following, we capture this curve in another curve fitting process. As a result, our model is extended with the

3.2 An Empirical Model for Response Time Prediction

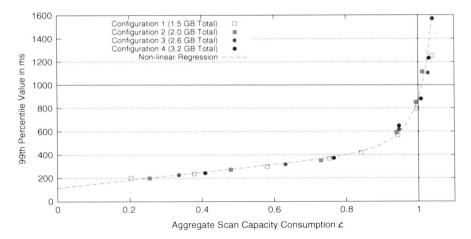

Fig. 3.4 Scan capacity consumption for multiple tenant configurations

Table 3.3 Coefficients for Eq. (3.6)

Coefficient	Value	95 % confidence bounds	
		Lower	Upper
p_1	334.4	−654.8	1,324
p_2	82.66	−9,748	9,913
p_3	1.833	−98.95	102.6
p_4	9.875	−342	361.8
p_5	33.46	−9,972	10,040

ability to predict the 99-th percentile value produced by a server given an arbitrary set of tenants for which request rates and sizes are known.

The data points in Fig. 3.4 suggest a linear increase of the 99-th percentile value up to $\mathcal{L} \approx 0.8$. Afterwards, the 99-th percentile value increases exponentially due to contention of the available bandwidth between CPU and main-memory. Based on this observation we can describe a function connecting all data points in Fig. 3.4 as a linear combination of a linear and an exponential function, as shown in Eq. (3.6):

$$f(\mathcal{L}) = p_1 \cdot \mathcal{L} + p_2 \cdot \exp(p_3 \cdot \mathcal{L}^{p_4}) + p_5 \qquad (3.6)$$

The coefficients in Eq. (3.6) can be determined using non-linear regression. In this case, a trust-region method [86] has been used and the values for the coefficients were determined as shown in Table 3.3.

The fit goodness is shown in Table 3.4. The RMSE of 206.5 ms seems high, given that our interest lies particularly in predicting 99-th percentile response times up to 1,000 ms.

Figure 3.5 shows a Q-Q plot comparing a sample of 99-th percentile response times that have been obtained in actual experiments to the 99-th percentile values

Table 3.4 Goodness of fit for Eq. (3.6)

Sum of squares due to error (SSE)	4.266×10^5
R^2 (coefficient of determination)	0.8979
Adjusted $\bar{R}^2$	0.8571
Root mean square error (RMSE)	206.5
RMSE for values < 1,000 ms	31.2088

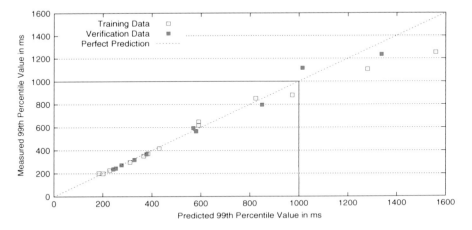

Fig. 3.5 Q-Q plot for estimated and measured 99-th percentile values

that have been predicted for these points using $\mathcal{L}$. Again, the tenants in the sample are heterogeneous. The sample is split into training data and verification data. The training data has been used to generate the model (i.e. for fitting the coefficients of Eq. (3.6)), while the verification data is solely used for quantifying the accuracy of the model. Both training and verification data are shown in Fig. 3.5. As can be seen in the figure, the prediction has a high accuracy for 99-th percentile values less than 1,000 ms, which is the range of particular interest to us. Larger 99-th percentile values result in a violation of our performance SLO regardless by how much the 99-th percentile value exceeds the 1,000 ms mark. When restricting the analysis of the goodness of the curve fitting result to data points with a 99-th percentile value smaller than 1,000 ms, the RMSE of the prediction is 31.2088, which is acceptable.

Although regression is a rather simple technique in comparison to more sophisticated techniques for response time prediction found in the literature (e.g. the machine-learning approach by Ganapathi et al. [47]), it proves to be capable of reliably predicting 99-th percentile values in our case. Given the simplicity of our model, it can easily be applied to settings with different server specifications and query profiles as long as the queries tend to be scan intensive, which is the case for both data warehouses and modern transactional systems based on a mixed workload.

3.3 Extending the Model with Batch Writes

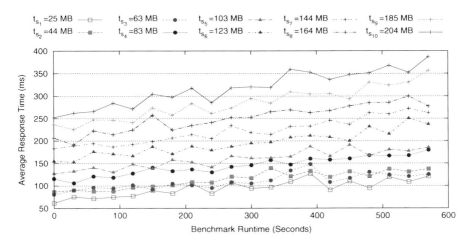

Fig. 3.6 Increasing response times due to periodic writes

3.3 Extending the Model with Batch Writes

Having established a basic model for predicting 99-th percentile values in the read-only case, we shall now investigate whether this model can be extended to include writes. In-memory column databases typically handle writes using a special auxiliary data structure. In SAP HANA, for example, each attribute of a table has a main store and a differential buffer. Both inserts and updates are appended to the differential buffer [71,93,103]. Both main store and differential buffer are dictionary compressed. However, the values in the differential buffer are not sorted to allow for fast appends. Read queries are executed against both main store and differential buffer, tuple validity and visibility is resolved inside the database engine, and the resulting rows are finally returned to the client. Figure 3.6 shows the impact of a growing differential buffer on read queries. As described in Sect. 3.1, we configured the benchmark to perform one write for each tenant every 5 min plus an initial offset to prevent write caravaning. Each write made a tenant's fact table grow by 0.05 % of its original size (i.e. the size at the beginning of the run). The response times shown in Fig. 3.6 averaged across all tenants of the same size and are grouped by tenant size. The cost for reading from the differential buffer increases linearly with its size. Consequently, main store and differential buffer must be merged occasionally (see also [71]), which is resource intensive but in effect lowers query response times as the differential buffer is cleared. Merge operations do not occur during the following experiment.

Figure 3.7 shows the response times in the 99-th percentile that were measured for different sets of heterogeneous tenants and varying values for $\mathcal{L}$, similar to the experiment in the previous section, but this time including periodic writes. For comparison, the data points from the previous experiment (which did not

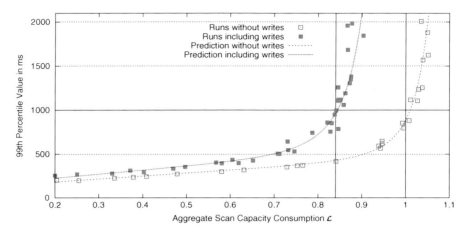

Fig. 3.7 Capacity of a single instance with batch writes

Table 3.5 Coefficients for Eq. (3.6) with writes

Coefficient	Value	95 % confidence bounds	
		Lower	Upper
p_1	457.9	−736.2	1,652
p_2	204	−7,152	7,560
p_3	5.609	−17.33	28.55
p_4	9.084	−94.36	112.5
p_5	−70.1	−7,762	7,621

contain writes) are also shown in the figure. As a consequence of the batch writes, the capacity of the server is saturated at a lower read request rate level. The slope of the curve begins to grow exponentially already at a lower value for $\mathcal{L}$ in the presence of writes. The maximum possible value of $\mathcal{L}$ on a server with concurrent batch writes is reduced to $\mathcal{L} \approx 0.84$. Again, the data points on the curve have been used in a curve fitting process. To do so, a function of the same general form as in the read-only experiment was used (cf. Eq. (3.6)).

The coefficients for Eq. (3.6) including batch writes are given in Table 3.5. Again, the coefficients have been estimated using a trust-region algorithm.

Table 3.6 shows an evaluation of the fit goodness of Eq. (3.6) and the coefficients in Table 3.5. Given the relatively high RMSE of 178.2 ms across all data points, we again calculated the RMSE for the relevant range of our predictions, which is 99-th percentile values less than 1,000 ms. The RMSE for this subset of the data is 62.3592.

We have extended our model to capture the drop in capacity available for processing read requests as a consequence of batch transactions. Note that this experiment and the curve fitting process are highly specific to (i) the write rate that we have chosen and (ii) the average growth of a tenant's differential buffer

Table 3.6 Goodness of fit for Eq. (3.6) with writes

Sum of squares due to error (SSE)	1.175×10^6
R^2 (coefficient of determination)	0.9313
Adjusted $\bar{R}^2$	0.9238
Root mean square error (RMSE)	178.2
RMSE for values <1,000 ms	62.3592

as a consequence of a write. As these parameters are varied, the function shown in Fig. 3.7 will move further to the left or the right of the chart.

3.4 Extending the Model with Migrations

As a SaaS provider, the ability to migrate tenants between servers without downtime is of vital importance, since service unavailability is one of the main drivers for customer churn. One example situation that necessitates migration is reacting to changes in the load pattern of one or more servers in the cluster. Such changes, typically driven by increasing request rates of individual tenants, might require re-balancing the assignment of tenants to servers. However, similar to batch writes, the migration of tenants between servers consumes resources that would otherwise be available for query processing. Especially when migration is being performed in response to an overload situation on a server, SLO violations might occur as a direct consequence of the additional load put on the server by migrating tenants away from it. In the following, we study the extent to which migration impacts the capacity of a server, and for how long. Our hypothesis is that, similar as for batch writes, migration reduces the maximum value of $\mathcal{L}$ by a fixed amount, regardless of the size of the tenant being migrated. Tenant size should only affect the duration of the migration. During migration, the source and destination server have different tasks: apart from the packing and unpacking of tenant data handled by both servers, only the destination server has to preload the data into main memory as part of the migration process. Data transfer again affects both servers involved. We thus analyze the capacity drop during migration separately for source and destination servers.

3.4.1 Resource Consumption of Migrations

In the following experiment, a single tenant is repeatedly migrated from one server to another. To perform a migration, the tenant data is packed on the source server, transferred to the destination server, unpacked on the destination server, and loaded into memory on the destination server. At this point, query processing for the tenant being migrated switches from the source to the destination server. In a production system, a complete migration would also entail deleting the tenant on the

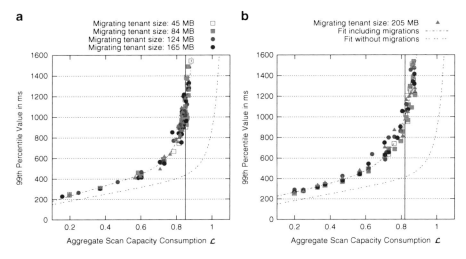

Fig. 3.8 99-th percentile values with ongoing migrations. (**a**) By bandwidth consumption and various migration sizes (Source server). (**b**) By bandwidth consumption and various migration sizes (Destination server)

source server. In this experiment, the tenant is not deleted but instead the migration process is instantly repeated. Our measurements show that deleting a tenant does not have a measurable impact on ongoing queries.

The servers involved in migrations were filled up with heterogeneous tenants of various data set sizes and request rates. In addition to continuously migrating a tenant away from the server, the server was exposed to a constant read workload. The aggregate value of $\mathcal{L}$ on the server was varied, similar to the previous experiments. The experiment is repeated several times so that tenants of different sizes are migrated. Figure 3.8 shows the response times in the 99-th percentile measured at different values of $\mathcal{L}$. Figure 3.8a focuses solely on the source server of the migration, whereas Fig. 3.8b shows the situation on the destination server. For comparison, the 99-th percentile values for the normal case (where no migrations occur) are also shown in both charts.

We can draw three observations from the charts in Fig. 3.8:

1. While involved in a migration, a server can sustain a lower value of $\mathcal{L}$ before violating the SLO. With ongoing migrations, the capacity drops to $\mathcal{L} \approx 0.85$ on the source server and $\mathcal{L} \approx 0.82$ on the destination server. These barriers are indicated using vertical bars in Fig. 3.8.
2. Again, the data points align on a curve (both in the source and destination server case). Again, this curve increases linearly at first and then grows exponentially as $\mathcal{L}$ increases. We can thus reuse Eq. (3.6) once more as the basis for fitting a function through the data points. The coefficients obtained in the curve fitting process for source and destination servers are shown in Table 3.7a and 3.7b, respectively.

3.4 Extending the Model with Migrations

Table 3.7 Coefficients for Eq. (3.6) with migrations

Coefficient	Value	95 % bounds	
		Lower	Upper
(a) Source server			
p_1	501.9	−696.8	1,701
p_2	262.5	−8,372	8,897
p_3	4.697	−21.87	31.26
p_4	8.963	−83.54	101.5
p_5	−130.4	−9,067	8,807
(b) Destination server			
p_1	380.9	−202.7	964.5
p_2	11.31	−269.3	292
p_3	6.001	−17.4	29.4
p_4	2.282	−9.427	13.99
p_5	177	−2.45	356.5

Table 3.8 Goodness of fit for Eq. (3.6) with migrations

	Source	Destination
Sum of squares due to error (SSE)	2.314×10^6	1.911×10^5
R^2 (coefficient of determination)	0.9209	0.9767
Adjusted $\bar{R}^2$	0.9157	0.9746
Root mean square error (RMSE)	194.8	65.9
RMSE for values <1,000 ms	67.3171	77.6308

3. The size of the tenant being migrated appears to be negligible as far as the alignment of the data points on a curve is concerned.

These observations verify our hypothesis that, similar to batch writes, migration reduces the capacity of a server by a fixed amount, regardless of the size of the tenant being migrated.

The fit goodness for function $\mathcal{L}$ in the presence of migrations is shown in Table 3.8. The $\bar{R}^2$ values indicates a good fit but the RMSE of 194.4 for the source server is high. We thus investigate the accuracy of the model including migrations using Q-Q plots.

Figure 3.9 shows the predicted and measured 99-th percentile values for source and destination servers, respectively. As is the case without migrations, the quality of the prediction is very precise up to a value of 600 ms. Afterwards, when the function begins to grow exponentially, the prediction becomes less accurate. However, the RMSE for all 99-th percentile values in the relevant range up to 1,000 ms calculated on the verification data is 67.3171 ms for the source server and 77.6308 ms for the destination server.

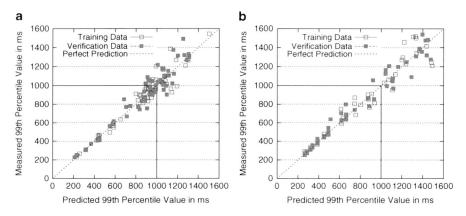

Fig. 3.9 Prediction of 99-th percentile values during migrations. (**a**) Source server. (**b**) Destination server

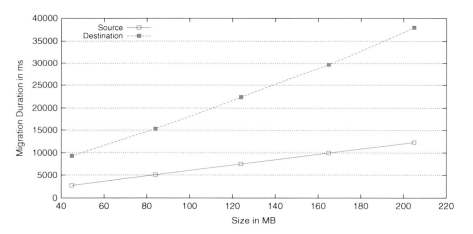

Fig. 3.10 Duration of migration impact by size of migrated tenant

3.4.2 Duration of Migration

We investigate for how long a migration reduces the maximum value of $\mathcal{L}$ before our performance SLO is violated. Again, we analyze source and destination servers of a migration separately. In the following experiment, the load on both source and destination servers constant at $\mathcal{L} \approx 0.605$. Figure 3.10 shows the duration of a migration for source and destination servers depending on the size of the tenant being migrated. The duration on the destination servers is higher than on the source servers because more processing is required. A source server only has to pack tenant data and transfer it to the destination server, whereas a destination server has to receive the data, unpack and import it, and load it into memory. The last step—preloading columns into memory—is particularly expensive and accounts for the largest fraction of the work during the migration process.

We observe that the duration of a migration increases linearly with the size of the tenant being migrated. This result shows that there is very little fixed overhead in performing a migration. The cost of migrating two smaller tenants is thus the same as for migrating a larger tenant of the same total size.

3.5 The Impact of Virtualization

The experimental results from which we built the empirical models presented in this section were obtained on Amazon EC2 [6], i.e. on virtual machines. As we have seen in Sect. 2.1, an important factor for achieving good scan performance in in-memory column databases is that few cache misses are produced during query processing. One reason for the low number of cache misses is that the hardware prefetcher of the CPU is often able to fill the L2 cache before the data is being read. The purpose of this section is to show that running the database inside a virtual machine has no detrimental effect on these aspects, and thus on performance. To do so, we investigate the performance of a scan-intensive workload both in non-virtualized and a virtualized environment.

In the following, we present a simple experiment, where we compare an installation of SAP HANA on an Intel Xeon E5450 server that runs Linux natively to an installation of HANA inside a Xen [25] virtual machine on the same server. The virtual machine was configured such that it can use all the resources available on the physical system (i.e. CPU and main memory). We ran the SSB-MT workload with one single tenant in the native installation and increased the number of parallel users for our tenant from 1 to 12. Figure 3.11 shows that the throughput in terms of queries per second is 7 % lower on average in the Xen-virtualized environment, and that both configurations exhibit a similar behavior otherwise. We believe that this overhead is largely due to the fact that the operators in HANA write intermediate results to main memory during query processing. For writing, virtual memory pages must be physically allocated to the HANA process, which is handled inside a page fault exception. Page faults are slightly more than twice as expensive inside a Xen virtual machine as in native Linux, since the corresponding system call is handled by the hypervisor and not the operating system kernel, which adds a level of indirection [85].

Minhas et al. have conducted a similar study using PostgresSQL and the TPC-H benchmark [85]. The average slowdown of the TPC-H queries is reported to be 9.8 %, which is consistent with our results. In [55] we present more experiments concerned with database performance in virtualized environments.

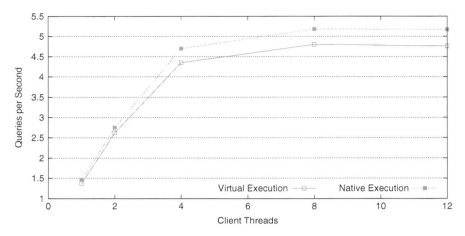

Fig. 3.11 Overhead of execution in a virtualized environment

3.6 Remarks

In this chapter, we have developed an experimental model for predicting a server's response time in the 99-th percentile in the presence of multi-tenancy. The inputs to this model are the sizes and request rates of the tenants to be assigned to a server. Tenants can be completely heterogeneous in terms of size and request rate. The input data is fed into a function $\mathcal{L}$, which, as output, returns the 99-th percentile response time of the server over a 10 min sliding window. With a mean error of 31 ms, the prediction process provides a high accuracy. Despite this, there are several limitations of our approach. In the following, we discuss these limitations and how they could possibly be addressed.

The parameter values obtained using regression are specific to the database schema and workload used during experimentation. In our case, the SSB benchmark was used. If one were to use different benchmark, such as TPC-H, or a custom application, then the experiments would have to be repeated to obtain new parameter values for the model. The methodology remains the same, however, as long as the workload has a scan-intensive profile. This is the case for a large fraction of applications, such as analytical applications, transactional applications with ad-hoc reporting features, or scan-intensive transactional applications (e.g. payment and dunning). Also, rather than performing frequent commits, the workload must write to the database in batches. Again, this is typically the case for analytical applications. It is also the case for SAP R/3, where the application server typically writes to the database in a batch process which is decoupled from the end-user transactions. Another limitation is that, while tenants can be heterogeneous in terms of size and request rate, all tenants must run the same workload. In a SaaS setting, however, it is reasonable to assume that tenants have a homogeneous workload.

3.6 Remarks

Our methodology is specific to in-memory column databases. Lang et al. [76] also benchmark heterogeneous tenants in the presence of multi-tenancy. Their study is based on Microsoft SQL Server and the TPC-C [112] benchmark. In their case, as long as all tenants fit into memory, the maximum number of queries processed per tenant before SLO violations occur remains constant even for varying tenant sizes. In our case, fewer queries can be processed as tenant size increases. The reason for this difference lies in the workload profile of TPC-C, which consists of point queries rather than scan-intensive queries. In the study of Lang et al., performance deteriorates when the size of all tenants exceeds the working set size of the SQL server process. The high variance in hard disk latency makes response times very difficult to predict [32]. The use of solid-state drives only slightly improves this situation [76].

Similar to changing the workload, the parameters of the model must be obtained anew when changing the hardware on which the workload is run. Again, the methodology remains the same. The experiments in this chapter were conducted on systems with a front-side bus-based memory architecture. Experimental evidence with newer, NUMA-based systems suggests that, for scan-intensive workloads, the amount of data moved from main memory to the CPU is still the limiting factor. Given the approximately four-fold increase in memory bandwidth of NUMA-based systems over front-side bus-based architectures (for so-called local memory), the amount of parallel queries that can be processed before the "scan capacity" is exhausted is higher, but the principles remain the same. Note that scan-intensive workloads are also CPU-intensive, due to the unpacking of the bit-compressed values in the columns and the aggregation operators executed during a scan. The data transfer between main memory and the CPU cache is the dominant cost factor, however.

Although many of the above cases require that the parameters of the model are re-calibrated, this process is actually easy given the simplicity of our model, which is based on statistical regression. In some cases, obtaining the model parameters might even be automated. We consider this a strong point of our methodology. Related approaches for performance modeling, such as the machine-learning-based approach by Ganapathi et al. [47], rely on complex mathematical procedures such as kernel methods [102].

Another strong point of our methodology is its extensibility. We have shown that our basic model can be re-used with varying parameters as disturbing factors such as batch writes or migrations are added. An interesting avenue for future work would be to capture the loss in server capacity coming from merging the differential buffer into the main store.

Chapter 4
The Robust Tenant Placement and Migration Problem

In this chapter, we introduce the problem of assigning tenants to servers so that performance SLOs are enforced and server cost is minimized. We call this problem the *Robust Tenant Placement and Migration Problem (**RTP**)*.

Before introducing incremental RTP, we begin with formalizing the static variant of RTP, where no initial placement is given, and tenant copies can be freely assigned without taking previous assignments into account. Afterwards, we introduce the incremental variant of the problem, building upon the formalization of static RTP.

Definition 4.1 (Valid Placement). A valid tenant placement is an assignment of at least two copies of a given number of tenants to a number of (cloud) servers so that

- No server is overloaded in terms of any of its resources (e.g. CPU, memory, or disk I/O),
- No server contains more than one copy of any one tenant, and
- The failure of a single server does not cause overloading any other server.

We consider a server *overloaded* when one of its resources is utilized beyond its capacity limit. This includes additional load that is redirected to a server when another server has failed. Servers must reserve a certain amount of "headroom" for such cases. Further, when a server fails, it must be possible to migrate tenants without violating performance SLOs. As we have seen in Sect. 3.4.1, migration reduces the capacity for processing requests on both the source and target server of a migration.

A tenant t is characterized by its size $\sigma(t)$ (i.e. the amount of space each replica consumes in memory and/or on disk) and its load $\ell(t)$. We view $\ell(t)$ as a metric that combines all resources related to processing queries, such as CPU usage or bandwidth utilization along the memory hierarchy. We have developed such a metric for in-memory column databases in Chap. 3. Recently, various work has proposed similar metrics for other types of databases and workloads [32, 37, 76] (which will also be discussed in Chap. 7). Our formulation of RTP is independent of the chosen workload modeling technique. Therefore, in the following, we use the symbol ℓ to denote a generic load metric, which is not necessarily identical to the function ℓ that we have developed in Chap. 3, although our metric could be used as such.

For the formalization of RTP, we assume an in-memory column database [42] and an enterprise mixed workload [73], such as encountered with SAP's on-demand applications [97]. This allows us to take advantage of two characteristics: (i) resource usage across multiple tenants is mostly additive for in-memory databases (as we have seen in the previous chapter) and (ii) load can be shared across all replicas of a tenant. Section 4.4 will discuss how these assumptions can be relaxed for other workloads and requirements.

At first glance, RTP resembles the two-dimensional bin-packing with conflicts problem [39], which, in addition to common bin-packing [77], has the notion of a so-called conflict graph used for specifying conflicts between items. Two conflicting items must not be assigned to the same bin. When adapting this to our case, conflicts would arise from the constraint that no server must hold more than one copy of the same tenant. However, RTP is different to the two-dimensional bin-packing with conflicts problem because our placements shall be *robust* towards server failures. A server failure causes an increase in load on those servers which also hold copies of the tenants on the failed server. RTP requires that, even when the additional load caused by a failure is included, no server becomes overloaded.

4.1 Static Placement

In the following, we use a so-called assignment formulation [63] to model and formalize RTP. We then discuss how multiple replicas of the tenants can be interleaved across servers to minimize cost and increase query throughput. Afterwards, we discuss two extensions of RTP: dynamically determining the number of replicas per tenant and ensuring that tenants can be migrated between servers without SLO violations.

A valid instance of RTP has the following data as **input**:

- $T \subseteq \mathbb{N}$, the set of tenants.
- $N \subseteq \mathbb{N}$, the set of available servers.
- $R = \{1, 2, \ldots, r(t)\}$, the replicas per tenant where $r(t) \geq 2$ is the (fixed) number of replicas per tenant; Sect. 4.1.2 contains details on how to obtain $r(t)$.
- $\sigma : T \to \mathbb{N}^+$, a function returning the DRAM requirement of a given tenant.
- $\text{cap}_\sigma : N \to \mathbb{N}^+$, a function returning the DRAM capacity of a given server.
- $\ell : T \to \mathbb{Q}^+$, a function returning the current load of a given tenant.
- $\text{cap}_\ell : N \to \mathbb{Q}^+$, a function returning the request processing capacity of a given server.

4.1 Static Placement

A valid solution of RTP must assign appropriate values to the following variables as **output**:

- A binary decision variable $y \in \{0,1\}^{N \times T \times R}$, where

$$y_{t,i}^{(k)} = \begin{cases} 1, & \text{if copy } k \text{ of tenant } t \text{ is on server } i \\ 0, & \text{otherwise} \end{cases}$$

- A binary decision variable $s \in \{0,1\}^N$, where $s_i = 1$ denotes that server i is active and otherwise, the server is not active.
- $p \in \mathbb{Q}_+^N$, where p_i denotes the fraction of the capacity of server i that must be left unused such that additional load due to a single server failure does not cause an SLO violation. We call p_i the *penalty* that must be reserved as headroom on server i.

Note that there are many cases that are *infeasible*, i.e. no solution can be found. This occurs, for example, when the set of available servers (input parameter N) has a cardinality of one.

The objective of RTP is to minimize the number of active servers, i.e.

$$\min \sum_{i \in N} s_i, \qquad (4.1)$$

subject to the following constraints. Constraint (4.2) ensures that each replica $1 \le k \le r(t)$ of a tenant t is assigned to a server exactly once.

$$\sum_{i \in N} y_{t,i}^{(k)} = 1 \qquad \forall t \in T, \forall k \in R \quad (4.2)$$

Constraint (4.3) ensures that no two copies of the same tenant are placed on the same server.

$$\sum_{k \in R} y_{t,i}^{(k)} \le 1 \qquad \forall t \in T, \forall i \in N \quad (4.3)$$

Constraint (4.4) ensures that the total size of all tenants on a server does not exceed the server's DRAM capacity. If at least one tenant is assigned to the server, s_i is set to one.

$$\sum_{t \in T} \sum_{k \in R} \sigma(t) \cdot y_{t,i}^{(k)} \le \text{cap}_\sigma(i) \cdot s_i \qquad \forall i \in N \quad (4.4)$$

Similarly, Constraint (4.5) ensures that the total load of all tenants on a server does not exceed the processing capabilities of the server. Due to our assumption that load can be shared across replicas, each server holding a replica of tenant t receives only $1/r(t)$th of $\ell(t)$.

$$\sum_{t \in T} \sum_{k \in R} \frac{\ell(t)}{r(t)} \cdot y_{t,i}^{(k)} + p_i \leq \mathrm{cap}_\ell(i) \cdot s_i \qquad \forall i \in N \quad (4.5)$$

Each server must be capable of handling potential additional load in case another server fails. The spare capacity reserved for this excess load is denoted by penalty p_i in Constraint (4.5). The following constraint defines the penalty.

$$p_i = \max_{j \in N: j \neq i} \sum_{t \in T} \sum_{k \in R} \sum_{k' \in R} \frac{\ell(t)}{r(t)^2 - r(t)} \cdot y_{t,i}^{(k)} \cdot y_{t,j}^{(k')} \qquad \forall i \in N \quad (4.6)$$

What fraction of a tenant's load must be added to p_i depends on the total number of replicas, or, more accurately, the number of remaining replicas after one server has failed. If server j handled a fraction $\ell(t)/r(t)$ of the load of tenant t load prior to the failure, then the remaining $r(t) - 1$ replicas of tenant t must share the load after the failure. Hence, the extra load that server i must support is:

$$\frac{\ell(t)}{r(t)} \frac{1}{r(t) - 1} = \frac{\ell(t)}{r(t)^2 - r(t)} \qquad (4.7)$$

Constraint (4.6) ensures that p_i is set large enough to guarantee that even the failure of the "worst case" other server $j \neq i$ would not result in overloading server i. The maximum function in Constraint (4.6) enforces this guarantee is given w. r. t. the worst case server. The constraint has a special property that renders standard heuristics for bin-packing unusable for RTP: given three servers U, V, and W, moving a tenant from V to W may increase p_U and thus render server U unable to sustain the extra load coming from another server failing.

4.1.1 Interleaving Tenant Replicas

Constraint (4.3) ensures that not more than one copy per tenant is assigned to a server. Beyond that, RTP does not restrict how tenant replicas should be laid out across servers. In this section, we revisit interleaved replica placement (first proposed by Teradata [57]) in the context of RTP. We present three examples showing that interleaving helps with (i) reducing the number of servers, (ii) increasing the query throughput of the cluster, and (iii) spreading out excess

4.1 Static Placement

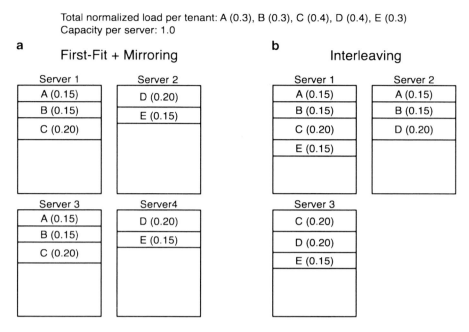

Fig. 4.1 Comparison of placement strategies with evenly distributed load among the replicas

load in the case of failures. Finally, we discuss the differences between mirroring and interleaving w.r.t. the probability of tenants becoming unavailable in case of multiple simultaneous server failures.

Example 4.1 (Interleaving Reduces the Number of Active Servers). As far as replication is concerned, it is state of the art to *mirror* tenant assignments [125]. Figure 4.1a shows a mirrored placement for five tenants A, B, C, D, and E with different resource requirements, shown in parentheses. We do not consider tenant size in this example. The mirrored placement has been obtained using Johnson's well-known first-fit algorithm [62] as follows: the tenants are first sorted by load in decreasing order. They are then assigned one by one to the first server until the server is full. The remaining tenants are assigned to the second server, and so forth. The resulting placement is finally mirrored, doubling the number of servers from 2 to 4. This approach has the disadvantage that the cluster must be substantially over-provisioned as headroom for additional load from server failures must be reserved on each server. This leads to a situation where the servers are under-utilized by more than 50% during normal operations. Note that the load is distributed evenly across all copies of a tenant (e.g. the total load of tenant A of 0.3 is spread across *server 1* and *3* in Fig. 4.1a). Upon the failure of a server, its mirror must take over the entire load.

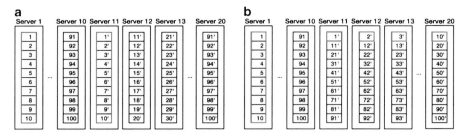

Fig. 4.2 Mirrored vs. interleaved placement in Example 4.2. (**a**) Mirroring . (**b**) Three replicas per tenant

In order to minimize the number of required servers, we use a technique called *interleaving* that tries to avoid co-locating any pair of tenants on more than one server. One advantage in comparison to mirroring is that, in case of a server failure, the excess load spreads across multiple nodes allowing to decrease the required "headroom" to handle failures and thus, improve utilization. Figure 4.1b shows an interleaved placement for the same five tenants as in the mirroring example. In the interleaved placement, a failure of *server 1* would redistribute the load of tenants *A* and *B* to *server 2* and the load of tenants *C* and *E* to *server 3*. As a result, the placement requires only three instead of four servers while no server becomes overloaded in case any server fails.

Example 4.2 (Interleaving Increases Query Throughput). In this example we compare a mirrored and an interleaved placement both using the same number of servers. We show that interleaved replica placement can have a positive impact on the maximum query throughput that a cluster of database servers can achieve. Consider a setup with 100 tenants on Amazon EC2 using the Rock clustering infrastructure described in Sect. 2.5. All tenants are homogeneous, i.e. they have exactly the same size (i.e. six million rows in the fact table) and the same load. We assign ten tenants to each server. There are two copies per tenant, hence 20 servers in total. The resulting mirrored and interleaved placements are shown in Fig. 4.2. In the interleaved placement, no two tenant replicas appear together on more than one server. We run both placement configurations under normal conditions as well as in the presence of failures. In the failure case, every 60 s one randomly chosen server is manually failed. In this particular setup, we observed that servers have a mean time to recover of 30 s, thus, on average, one out of 20 servers is unavailable for approximately 50 % of the benchmark period. Note that this is a very high failure rate which is unlikely to occur in practice. However, this is an interesting edge case that brings out the differences between mirroring and interleaving when operating under failures.

We simultaneously increase the base load of all tenants (by adding more concurrent users) until the first server in the cluster violates the performance SLO (i.e. response times in the 99-th percentile exceed 1 s). Table 4.1 shows the result

4.1 Static Placement

Table 4.1 Maximum number of concurrent users before SLO violation occurs

	Mirroring	Interleaving	Improvement
Normal operations	4,218 users	4,506 users	7 %
Periodical failures	2,265 users	4,250 users	88 %

of the experiment: under normal operating conditions, interleaving achieves a 7 % higher throughput than mirroring. The reason for this difference is that—at a large number of concurrently active users—an individual tenant t may receive a relatively high number of simultaneous queries, which creates temporary hotspots on the two servers of t. The interleaved configuration is better at spreading out these short-lived load spikes in the cluster than mirroring. In the presence of periodical failures, the maximum throughput that the mirrored configuration can sustain before SLO violations occur drops by almost 50 % when compared to normal operations, as expected. Interleaving, in contrast, completely hides the failure from a throughput perspective. Interestingly, even in the presence of failures, the interleaved configuration can support 32 more active user sessions than the mirrored configuration during normal operating conditions.

Example 4.3 (Interleaving Spreads out Excess Load Across the Cluster). Based on the experiment presented in the previous example, we analyze the impact of server failures over time for both mirroring and interleaving. The operating point for this example is 2,265 simultaneous users, i.e. where SLO violations begin to occur for the mirrored configuration when failures are injected into the cluster (cf. Table 4.1). We analyze the total load including penalty $\sum_{t \in T} \sum_{k \in R} \frac{\ell(t)}{r(t)} \cdot y_{t,j}^{(k)} + p_j$ (cf. Constraint (4.5)) on the *worst* server j in the cluster. At each point in time, the particular server with the highest load including penalty among all servers is chosen as the worst server. Thus, multiple servers fill the role of the worse server during the course of the benchmark run. Figure 4.3 compares load plus penalty predicted using the model developed in Sect. 3.2 to the actually *measured* load on the worst server.[1] The arrows in Fig. 4.3 denote the points during benchmark execution where failures are injected into the cluster. The figure shows that interleaving achieves a much better redistribution of the excess load incurred by failures than mirroring, since the load including penalty on the worst server is around 0.6 for interleaving and 1.0 for mirroring. Also, the load measured on the worst server in the mirrored variant varies considerably both when it is hit with the excess load coming from the failure and when the excess load subsides after the failed server is back on-line after recovery. For the interleaved variant, the measured load varies much less, since the excess load is shared among ten servers, as opposed to one server in the mirrored placement. Hence, the excess load *per server* is much lower in the interleaved configuration than in the mirrored one.

---

[1] Technically, the response time in the 99-th percentile has been measured and transformed into load using Eq. (3.6).

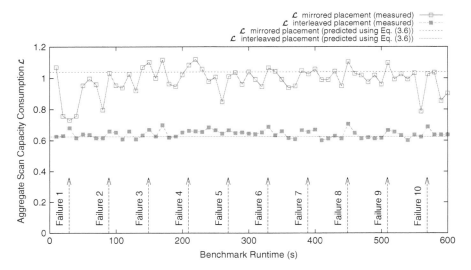

Fig. 4.3 Aggregate scan capacity consumption in the presence of failures for mirrored and interleaved placement on the worst server

We observe that the deviation between measured and predicted load is higher in the mirrored case. The reasons are that (i) the prediction does not include how load on the worst server ramps up and down as servers fail and recover and (ii) the predictor developed in Sect. 3.2 becomes less accurate for 99-th percentile values greater than 1,000 ms, which is the operating point of the mirrored setup in this experiment.

Risk of Tenant Unavailability in Case of Multiple Simultaneous Server Failures

In the following, we briefly discuss the risk of tenants becoming unavailable in the presence of two simultaneous server failures. For mirroring, intuition suggests that the probability of two servers failing at the same time, where one server is the mirror server of the other server, is low. Yet, the impact of this event is high since many tenants become unavailable at once. For interleaving, the intuition is that for any two servers that fail simultaneously, there is a high probability of individual tenants becoming unavailable. The number of affected tenants is expected to be lower than in the mirrored case; the probability of a tenant becoming unavailable, however, should be higher. The question is for which of both cases the product of probability and impact is greater, or, in other words, which one is the riskier placement strategy.

Let us consider a situation with $|T|$ tenants with two replicas each and n servers, two of which fail simultaneously. From the perspective of the customer (i.e. an individual tenant t), it does not matter whether a mirrored or interleaved assignment

4.1 Static Placement

is chosen. The probability of t being assigned to both failed servers is the same, both in the mirrored and the interleaved case. The perspective of the server provider is less obvious. For quantifying the provider's risk, we assume that a certain cost $c(t)$ is associated with a tenant becoming unavailable. Equation (4.8) shows the expected costs for the service provider in the mirrored case:

$$q \cdot \frac{1}{n-1} q \cdot \sum_{t}^{T_{\text{fail}}} c(t) \tag{4.8}$$

In Eq. (4.8), q denotes the probability of a server failure and T_{fail} is the set of tenants on the first server that fails. After an arbitrary server has failed first, the probability that its mirror server fails is $\frac{q}{n-1}$ since there are $n-1$ remaining servers after the first failure. The cost is simply aggregated across all failed tenants. Equation (4.9) shows the expected costs in the interleaved case:

$$q \cdot q \cdot \sum_{t}^{T_{\text{fail}}} \frac{1}{n-1} c(t) \tag{4.9}$$

In the interleaved case, the probability that two arbitrary servers fail is $q \cdot q$. Then, for each a tenant from the set of failed tenants on the first server, the probability of having another copy on the second failed server is $\frac{1}{n-1}$.

Proposition 4.1. *From a provider perspective, the risk of tenants becoming unavailable is the same for both mirrored and interleaved replica placement strategies.*

Proof. Follows directly from the equality of Eqs. (4.8) and (4.9). □

4.1.2 Choosing the Number of Replicas

In the previous sections, the number of replicas per tenant $r(t)$ was treated as an input parameter to our optimization problem. In the following, we discuss how to compute $r(t)$, for example in a preprocessing step. Intuition suggests to set $r(t)$ as low as possible, since (i) more replicas require more space, which should result in a higher number of active servers, and (ii) the problem becomes more constrained since more instances of Constraint (4.3) are being generated. Still, increasing the number of replicas beyond $r(t) = 2$ becomes necessary when the load of a tenant is so high that a single server cannot handle half of it. In this case, our assumption that load is shared among replicas allows for *scaling out* across multiple servers (by increasing the number of replicas). The number of copies $r(t)$ of a tenant t must be chosen such that $\ell(t)/r(t) < \text{cap}_\ell(i)$. In addition, server i must be able to handle the extra load coming from another server failing that also holds a copy of t. Hence, we must choose $r(t)$ in such a way that the following inequality applies:

$$\frac{\ell(t)}{r(t)} + \frac{\ell(t)}{r(t)^2 - r(t)} \leq \text{cap}_\ell(i) \qquad \forall i \in N \qquad (4.10)$$

We rearrange Eq. (4.10) for $r(t)$ as follows:

$$\frac{\ell(t)}{r(t)} + \frac{\ell(t)}{r(t)^2 - r(t)} \leq \text{cap}_\ell$$

$$\Leftrightarrow \quad \ell(t)r(t) - \ell(t) + \ell(t) \leq \text{cap}_\ell(r(t)^2 - r(t))$$

$$\Leftrightarrow \quad \ell(t)r(t) \leq \text{cap}_\ell r(t)^2 - \text{cap}_\ell r(t)$$

$$\Leftrightarrow \quad \ell(t) \leq \text{cap}_\ell r(t) - \text{cap}_\ell$$

$$\Leftrightarrow \quad \ell(t) + \text{cap}_\ell \leq \text{cap}_\ell r(t)$$

$$\Leftrightarrow \quad \frac{\ell(t) + \text{cap}_\ell}{\text{cap}_\ell} \leq r(t)$$

$$\Leftrightarrow \quad \frac{\ell(t)}{\text{cap}_\ell} + 1 \leq r(t)$$

Inequality (4.10) ensures that $r(t)$ is chosen such that a server is not overloaded from assigning a copy of t to the server. At the same time, we must ensure that each tenant has at least two replicas.

Definition 4.2 (Minimum Number of Replicas per Tenant). We define the lower bound on the number of replicas per tenant $r(t)$ as follows.

$$r(t) := \max\left(2, \left\lceil \frac{\ell(t)}{\text{cap}_\ell(t)} + 1 \right\rceil\right) \qquad (4.11)$$

The following example shows that—in contrast to our intuition—increasing the number of replicas beyond the lower bound can lead to placements with fewer servers.

Example 4.4 (Impact of the Number of Replicas on the Number of Servers). Consider four tenants A, B, C, and D, each with a load of $\ell = 1.0$ and servers with capacity of $\text{cap}_\ell = 1.0$. For the purpose of simplification, we do not consider the DRAM requirements σ of the tenants in this example. At two replicas per tenant, as shown in Fig. 4.4, eight servers are necessary for assigning all tenants to servers. The load on all servers including spare capacity reserved to accommodate potential server failures (i.e. p_i) is 1.0. If we allow three replicas per tenant, as shown in Fig. 4.4, then a total of six servers is sufficient. Also in this case, the load on all servers including p_i is 1.0.

4.1 Static Placement

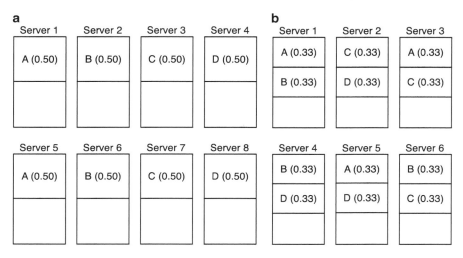

Fig. 4.4 Required number of servers dependent on $r(t)$. (**a**) Two replicas per tenant. (**b**) Three replicas per tenant

Example 4.4 suggests computing the number of copies per tenant dynamically as opposed to always using the lower bound as defined in Constraint (4.11). This does, unfortunately, drastically increase the space of possible solutions, which suggests to look for heuristic approaches for determining the number of replicas of each tenant. A detailed discussion on the complexity of RTP will be provided in Sect. 4.3.

4.1.3 Recoverable and Flexible Placements

In the previous sections, we have formulated RTP in a way that a placement shall be *robust* towards one server failure. When a server fails, however, it must be possible to recreate the lost tenant replicas residing on the failed server on other servers. For each affected tenant, this is done by identifying live servers also holding a copy of the tenant. As a consequence of Constraint (4.2) and Eq. (4.11) there is at least one such server. The tenant replica can then be copied from the identified server to a new server.[2]

Migrations consume resources on servers participating in a migration. This temporarily reduces a server's capacity for processing requests ($\text{cap}_\ell(i)$). By how much the capacity degrades during migration depends on the actual schema and workload. In Sect. 3.4.1 we established that—for the used in-memory database and

---

[2]Another conceivable option would be to copy the tenant from a shared filesystem. We focus on shared-nothing architectures in this dissertation (cf. Chap. 2).

workload—the overhead migrating a tenant is 15 % on the source server and 18 % at the destination server. We use these values in our experiments. For the formal exposition of RTP, we describe these values as input parameters. The "deterioration" factor for the source server of a migration is called μ, with $0 < \mu < 1$. Similarly, ν denotes the factor by which the capacity of a migration target server is reduced.

Definition 4.3 (Recoverability). We call a placement recoverable when it contains two servers for each tenant that have enough spare capacity to act as the source server of a migration. Formally, recoverability is ensured with Constraint (4.12).

$$\forall t \in T \; \exists i,j \in N, i \neq j :$$

$$\sum_{t' \in T} \sum_{k \in R} \frac{\ell(t')}{r(t)} \cdot y_{t',i}^{(k)} + p_i \leq \mu \cdot \mathrm{cap}_\ell(i) \cdot s_i \; \wedge$$

$$\sum_{t' \in T} \sum_{k \in R} \frac{\ell(t')}{r(t)} \cdot y_{t',j}^{(k)} + p_j \leq \mu \cdot \mathrm{cap}_\ell(i) \cdot s_j \tag{4.12}$$

Constraint (4.12) enforces that enough spare capacity for every possible tenant migration is reserved via $\mu \cdot \mathrm{cap}_\ell(i)$.

Constraint (4.12) might be regarded as overly pessimistic since it also guarantees that migrations can be performed without SLO violations when the failed server is a safe migration source for the tenants assigned to it. For some cases, it might be enough that each tenant has at least one replica on a server from which it can be migrated away without violating performance SLOs.

Definition 4.4 (Flexibility). We call a placement flexible when it contains at least one server for each tenant that has enough spare capacity to act as the source server of a migration, formalized as follows:

$$\forall t \in T \; \exists i \in N :$$

$$\sum_{t' \in T} \sum_{k \in R} \frac{\ell(t')}{r(t)} \cdot y_{t',i}^{(k)} + p_i \leq \mu \cdot \mathrm{cap}_\ell(i) \cdot s_i \tag{4.13}$$

Definition 4.5 (Safe migration source). Each server i that satisfies Constraint (4.13) for a given tenant t is called a safe migration source of t.

4.2 Incremental Placement

So far, we have not considered the case that tenant load continuously changes over time. In this case, placements must be found that take the status quo, i.e. the current assignment of tenants to servers into account. Simply computing a new solution to

4.2 Incremental Placement

the optimization problem from scratch using static RTP is not viable. The reason is that the transformation from the current assignment to the new placement is done by migrating individual tenants while the tenants are on-line. As we have seen in the previous section, migration temporarily reduces a server's capacity for processing requests. Also, only so much migration can occur within a limited amount of time.

RTP must thus be solved periodically using an existing placement as the starting point. We call this problem *incremental RTP* (as opposed to static RTP as described above). The length of the reassignment interval limits the amount of data migration and thus the amount of change that can be applied to the original placement. The size of a tenant dictates how long it takes to migrate the tenant (cf. Sect. 3.4.1). The amount of migration that is permissible in each step also depends on the extent to which migrations can be performed in parallel (i.e. when multiple server pairs are involved). We do not study this problem. Instead, we assume that a fixed amount of data can be migrated in each step. Later, in Chap. 6, we analyze the impact of varying the migration budget on the quality of the solutions of incremental RTP.

In addition to the input data for static RTP described in Sect. 4.1, the incremental version of the placement problem requires an existing placement $y'^{(k)}_{t,i}$ and a migration budget $\delta \in \mathbb{N}$ as input variables.

For incremental RTP, we introduce the following additional constraints: Constraint (4.14) expresses that all servers onto which tenants shall be migrated must have enough spare capacity to do so without violating performance SLOs.

$$\forall i \in N : i \text{ is migration target}$$

$$\sum_{t \in T} \sum_{k \in R} \frac{\ell(t)}{r(t)} \cdot y^{(k)}_{t,i} + p_i \leq \nu \cdot \text{cap}_\ell(i) \cdot s_i \tag{4.14}$$

Constraint (4.15) ensures that the migration budget δ is never violated. For notational convenience, we define $T_{\text{mig}} := \{t \in T : \text{a copy of } t \text{ was moved}\}$. This formulation implicitly takes the previous placement y' into account. For implementing T_{mig} in an algorithm further modeling details are necessary, which will be presented in Sect. 5.2.5.

$$\sum_{t \in T_{\text{mig}}} \sigma(t) \leq \delta \tag{4.15}$$

Constraint (4.16) states that for every tenant t that is being migrated, at least one server with a copy of t has enough spare capacity to act as a migration source for t without becoming overloaded as a consequence of participating in the migration. Note the similarity to Constraint (4.13). The difference between both constraints is as follows. While Constraint (4.13) ensures that a valid solution of RTP contains a safe migration source server for all tenants, Constraint (4.16) takes the previous assignment of tenants to servers y' into account but assumes that the load of the tenants might have changed since y' has been produced. Increases in tenant load can lead to situations where tenants that had a safe migration source at the time

y' was computed do no longer have a safe migration source after load changes for the tenants have been observed. These load changes must be taken into account when computing the new solution y.

$$\forall t \in T_{\text{mig}} \, \exists i \in N :$$

$$\sum_{t' \in T} \sum_{k \in R} \frac{\ell(t')}{r(t)} \cdot y_{t',i}^{(k)} + p_i \leq \mu \cdot \text{cap}_\ell(i) \cdot s_i \qquad (4.16)$$

This is a given for *flexible* placements. Constraint (4.16) is similar to but weaker than Constraint (4.13), since it encompasses only those tenants that shall be migrated in the transition from y' to y.

Constraints (4.15) and (4.16) may render RTP infeasible in case of extreme load change in comparison to the given placement. In such cases, it may occur that (i) no server can act as a safe migration source for a tenant or (ii) the migration budget is not large enough for repairing all overloaded servers. When an infeasibility occurs, it becomes necessary to (temporarily) tolerate SLO violations while restoring a valid and flexible placement. Besides temporarily dropping constraints, a change in the objective function becomes necessary to minimize SLO violations. Instead of minimizing the number of active servers, a placement shall be found with the lowest possible number of overloaded servers, which can be formalized as follows.

Definition 4.6 (Excess Load). Let $e \in \mathbb{Q}_+^N$ be the excess load on a server. For $i \in N$, e_i is defined as follows:

$$e_i := \sum_{t \in T} \sum_{k \in R} \frac{\ell(t)}{r(t)} \cdot y_{t,i}^{(k)} + p_i - \text{cap}_\ell(i) \qquad (4.17)$$

Two alternative objective functions for the optimization problem for cases of infeasibility as discussed above are

$$\min \sum_{i \in N} e_i \quad \text{or} \quad \min \max_{i \in N} e_i. \qquad (4.18)$$

We have formulated RTP as an assignment problem. Note that other formulations of RTP are possible, such as the representation of RTP as a mixed integer program. Choosing other decision variables similar to [117] is also conceivable. Our modeling decision in favor of the assignment formulation was motivated by its flexibility and expressiveness.

Note that static RTP is a special case of incremental RTP where (i) no initial placement y' is given, (ii) $\nu_i = \mu_i = 1$ for all $i \in N$, and (iii) $\delta = \infty$.

Note further that an optimal solution of static RTP is a lower bound for any optimal solution of incremental RTP.

4.3 Complexity Analysis

In this section, we show the $\mathcal{NP}$-completeness of both static and incremental RTP. To demonstrate that RTP is $\mathcal{NP}$-complete, the following two conditions must be satisfied:

1. RTP $\subseteq \mathcal{NP}$, and
2. RTP is $\mathcal{NP}$-hard, i.e. there is an $\mathcal{NP}$-complete problem that can be *reduced* to RTP.

Proposition 4.2. *RTP* $\subseteq \mathcal{NP}$.

Proof. RTP $\subseteq \mathcal{NP}$ can be shown by demonstrating that a solution to RTP can be verified in polynomial time in the size of the input instance. A verifier for solutions to RTP must validate the following properties of a solution:

1. Each tenant is assigned to exactly $r(t)$ different servers. This property can be checked by looping over all tenants, and for each tenant t again looping over all servers, counting the number of assignments of t. Let R be an instance of RTP with $|T|$ tenants and $|S|$ servers, and number of replicas $r(t)$ of the tenant t with the highest number of replicas is $r_{\max}$. Then, this step requires $r_{\max} \cdot |T| \cdot |N|$ operations.
2. No server is beyond its DRAM capacity cap_σ. This property can be checked by looping over all servers, and for each server looping over all tenants t assigned to this server, summing up the tenants' DRAM requirements $\sigma(t)$. This step requires $|N| \cdot r_{\max} \cdot |T|$ operations.
3. All servers have enough spare capacity so that no server becomes overloaded when any one other server fails. This property can be checked by looping over all servers, and for each server i again looping over all tenants t assigned to i, summing up the tenants' load requirements $\frac{\ell(t)}{r(t)}$, and computing the penalty p_i. The penalty of a server i can be computed by looping over all servers except i, and for each other server j looping over all tenants assigned to this server and summing up $\frac{\ell(t)}{r(t)^2 - r(t)}$ for all tenants that are assigned to both i and j. The penalty p_i is then the maximum value across all (i, j) pairs. Note that checking this property subsumes checking that no server is overloaded. This step requires $|N| \cdot r_{\max} \cdot |T| \cdot (|N| - 1)$ operations.
4. Each tenant has a safe migration source as specified in Definition 4.5. This property can be checked by looping over all tenants, and for each tenant looping over all servers that the tenant is assigned to, until a server is found with a total of load and penalty smaller than $\mu \cdot \text{cap}_\ell(i)$. This step requires $r_{\max} \cdot |T| \cdot |N|$ operations.

Since each check requires a polynomial amount of computation, the total amount of computation required for validating a solution of RTP is also polynomial. □

Theorem 4.1. *Static RTP is $\mathcal{NP}$-hard.*

Proof. For showing the $\mathcal{NP}$-hardness of static RTP, we demonstrate that the $\mathcal{NP}$-hard PARTITION problem [48] can be reduced to RTP in polynomial time. We show that, given an algorithm $\mathcal{A}$ for solving instances of RTP, we can use a many-one reduction from PARTITION to RTP to solve instances of PARTITION by

1. Transforming an instance of PARTITION into an instance of RTP,
2. Solving the instance of RTP using $\mathcal{A}$, and
3. Transforming the solution of RTP into a solution of PARTITION.

In the following, we describe polynomial-time transformations for step one and three.

A decision instance of the PARTITION is defined by a set of n elements with integer values $a_1, \ldots, a_n$. The decision question is whether the elements in the set can be partitioned into two disjoint subsets such that the sum of the values in both partitions is $(\sum_{i=1}^{n} a_i)/2$. Such instances can be transformed into a decision instance of RTP as follows: for each element i in the partition instance, a tenant t is created in RTP with $\sigma(t) = a_i$ and $\ell(t) = 0$. We assume that all servers have a DRAM capacity of $\text{cap}_\sigma = (\sum_{i=1}^{n} a_i)/2$ and a load capacity of $\text{cap}_\ell = 1.0$. Since all tenants have zero load, each tenant has two replicas, i.e. $r(t) = 2$. The decision question is whether a valid placement with four servers can be found. This decision instance of RTP can now be solved using $\mathcal{A}$. If no valid placement with four server can be found, we call the instance of RTP a *NO-instance*. A NO-instance of RTP induces a NO-instance of PARTITION.

For YES-instances of RTP, we have a solution with a set of servers $N = \{1, \ldots, 4\}$ and all tenants have a set of replicas $R = \{1, 2\}$. This solution consists of two pairs of two servers each such that both servers in each pair hold the same tenants. Our approach for mapping this solution to a solution of PARTITION entails (i) transforming the solution into a mirrored placement; (ii) deleting two out of the four servers in the mirrored placement; and (iii) mapping each of the two remaining servers to one subset of the solution to the original PARTITION instance. To realize the first step, we modify the solution of RTP such that all tenants on any one server have the same replica index, i.e. either $y_{t,i}^{(1)}$ or $y_{t,i}^{(2)}$ for all tenants t assigned to server i. In Definition 4.7, we define a function that we use to perform the necessary modifications.

Definition 4.7. We define a function $\text{swap}: (T, N, N) \rightarrow \{0, 1\}^{N \times T \times R}$, which swaps two replicas of the same tenant t on two server i and j.

$$\text{swap}(t, i, j) := \begin{cases} y_{t,i}^{(1)} \wedge y_{t,j}^{(2)}, & \text{if } y_{t,i}^{(2)} \wedge y_{t,j}^{(1)} \\ y_{t,i}^{(2)} \wedge y_{t,j}^{(1)}, & \text{if } y_{t,i}^{(1)} \wedge y_{t,j}^{(2)} \end{cases}, \quad t \in T, \ i, j \in N$$

Algorithm 1 shows how we transform the solution of RTP into a solution of PARTITION. The algorithm consists of three steps. In the first step, the solution is transformed into a mirrored placement by harmonizing all replica indexes using

4.3 Complexity Analysis

Algorithm 1 A polynomial-time algorithm for transforming solutions of RTP into solutions of PARTITION

▷ Step 1: Harmonize replica indexes
for server i in $N = \{1, \ldots, 4\}$ **do**
    for tenant t on i **do**
        if $y_{t,i}^{(2)} = 1$ **then**
            find other server j such that $y_{t,j}^{(1)}$
            call $swap(t, i, j)$
        end if
    end for
end for
▷ Step 2: Pick the right servers for deletion
if $\exists t$ such that $y_{t,2}^{(2)} = 1$ **then**
    if $\exists t$ such that $y_{t,3}^{(2)} = 1$ **then**
        delete servers $N = 1$ and $N = 4$
    else
        delete servers $N = 1$ and $N = 3$
    end if
else
    delete servers $N = 1$ and $N = 2$
end if
▷ Step 3: Map to solution of PARTITION
for server i in N **do**
    $S_i \leftarrow$ new empty subset of PARTITION solution
    for tenant t on i **do**
        $a_t \leftarrow \sigma(t)$, $a_t \in S_i$
    end for
end for
return S_1, S_2

---

the function *swap*. In the second step, two out of the four servers are deleted. The deleted servers are the mirror servers. In the third step, each of the two remaining servers are mapped to one of the two subsets that form the solution of the original PARTITION instance. It can easily be seen that Algorithm 1 is of polynomial time complexity. When following this procedure, a YES-instance of RTP induces a YES-instance of PARTITION. □

Note that Proposition 4.2 and Theorem 4.1 only demonstrate the *weak $\mathcal{NP}$-completeness* of static RTP, since the PARTITION problem, which is used for the reduction, is only weakly $\mathcal{NP}$-hard [48].

Corollary 4.1. *Incremental RTP is $\mathcal{NP}$-hard.*

Proof. Note that static RTP is a special case of incremental RTP where (i) no initial placement y' is given, (ii) $v_i = \mu_i = 1$ for all $i \in N$, and (iii) $\delta = \infty$. Therefore, incremental RTP is at least as hard as static RTP. □

4.4 Remarks

In our formulation of RTP, we assumed that the underlying database is an in-memory column database. This applies for two major reasons. Firstly, for this type of database, the main resource consumed by the system, DRAM bandwidth utilization, combines almost linearly across multiple tenants on the same physical machine. This simplifies the formal exposition of RTP since there is no inherited complexity from the underlying workload modeling approach. RTP can be extended for example for disk-based databases that require an elaborate model for disk I/O bandwidth utilization as described in [32, 76]. Such an extension could be done by adding a corresponding constraint similar to Constraint (4.4).

We further assumed a read-mostly workload. In our formalization, this assumption is manifested in Constraint (4.14), which states that load is distributed equally among a tenant's replicas. This constraint could easily be modified by splitting load into weighted read and write components, thereby making our formulation of RTP independent of the workload characteristics. In the case of a write-mostly workload, however, the load cannot be split among replicas; instead, all replicas are exposed to the full load (assuming that writes go to all replicas). Also, in that case, there is no redistribution of load in case of failure. All other aspects of the problem formulation, especially those related to on-line tenant migration, remain intact. An exhaustive study of the impact of writes on replicated tenant placement is beyond the scope of this dissertation. Here, we focus on mixed workload enterprise applications with a read-mostly workload [73] or analytical applications using an in-memory column database [42]. The log data used for our experimental evaluation (Chap. 6) was also taken from such an application.

For our experiments, we chose the load metric proposed in Chap. 3 of this dissertation. The primary reason is that our model includes migration cost, a fundamental prerequisite for RTP. The inclusion of other load metrics [32, 37, 76] is possible in principle.

Chapter 5
Algorithms for RTP

In this chapter, we present algorithms for solving RTP. Solutions must adhere to all constraints described in the previous chapter. Since it is not obvious what kind of algorithmic strategy (i.e. what algorithm *family*) is most promising for solving RTP, we develop solutions across the full spectrum of algorithm engineering: our approaches range from greedy heuristics, to metaheuristics involving randomness, and exact algorithms. Following the structure of Chap. 4, we begin with presenting our algorithms for static placement, forming the foundation for our incremental algorithms, which we discuss afterwards.

5.1 Algorithms for Static RTP

In the following, we introduce algorithms for static RTP. We begin with presenting our greedy heuristics and then proceed to other algorithms with increasing computational complexity.

5.1.1 Greedy Heuristics

Greedy heuristics are well-known to deliver good results for the related bin-packing problem. Therefore, it seems reasonable to also explore greedy heuristics for RTP. Another reason for considering greedy variants is that they are computationally less intensive than metaheuristics based on randomness and exact algorithms.

Our greedy algorithms are loosely based on the well-known best-fit algorithm [30], an improved variant of Johnson's first-fit algorithm [62]. When placing a single replica of a tenant, for each server its total load including its *penalty* (Sect. 4.1) is computed. We cache penalty on a per-server basis to speed up computation. The servers are then ordered according to load plus penalty in decreasing order. Similar to best-fit, the first server with enough free capacity is selected. If no active server has enough capacity, then the tenant is placed on a new server.

Apart from assigning tenants to servers based on the total load and the penalties of the servers, we also consider the other constraints of static RTP: for example, a tenant t cannot be added to a server i if, in consequence, the penalty of another server j would increase in a way that j becomes overloaded. Furthermore, a tenant t_1 cannot be placed on a server i if, consequently, another tenant t_2 loses its sole safe migration source server (cf. Definition 4.5). This occurs when the load on the target server i, after adding a replica of t_1, becomes larger than $\mu \cdot \text{cap}_\ell$ and the server previously in question was the only safe migration source of t_2 (or of any other tenant assigned to it). Naturally, a tenant cannot be added to a server if there is not enough available DRAM. This basic mechanism for placing a single replica of a tenant is called *robustfit-single-replica*. It is the basis for the algorithms robustfit-static-mirror and robustfit-static-interleaved, which will now be discussed.

robustfit-static-mirror begins with sorting all tenants by load (in descending order) and places the first replica of each tenant. Since there is no penalty when there is only one copy, the algorithm assumes a server capacity of $\frac{\mu \cdot \text{cap}_\ell(i)}{2}$ in this step, so that enough spare capacity is left unused to handle additional load in the case of server failures. Next, all servers are mirrored. Finally, the algorithm places additional replicas individually for tenants that require more than two replicas (see Sect. 4.1.2). robustfit-static-interleaved also sorts all tenants and then, tenant after tenant, places all replicas of each tenant. For the first replica of each tenant, a server capacity of $\mu \cdot \text{cap}_\ell(i)$ is assumed. This results in a placement where each tenant has a safe source server. For all further replicas beyond first copy, the algorithm assumes a capacity of $\text{cap}_\ell(i)$. Note that the latter does not result in any tenant losing its safe migration source server; this condition is checked for in robustfit-single-replica, the method for greedily placing individual tenant copies. robustfit-static-mirror naturally interleaves tenant replicas across servers.

Calculating the penalty for all servers is the most computationally intensive part of the two algorithms described above. As already discussed in Proposition 4.2, the penalty of a server i is computed by pair-wise for each pair (i, j) with $i \neq j$. Then, the maximum across all pairs is selected as the penalty for i. The complexity of computing the penalty of all server is thus quadratic in the number of servers $|N|$. Specifically, it has a complexity of $r_{\max} \cdot |T| \cdot |N| \cdot (|N| - 1)$ (with $|T| > |N|$). In addition to this complexity, the two algorithms described above need to sort all tenants ($|T| \log |T|$) and assign all replicas of each tenant $((|T|)^{r_{\max}})$. Their total computational complexity is thus proportional to $(|T|)^{r_{\max}+2} \cdot \log |T| \cdot |T| \cdot (|N|)^2$.

The next greedy algorithm for static RTP, robustfit-static-2atonce, is the only greedy algorithm that does not build upon the best-fit approach for placing individual replicas. Instead, for one tenant, it tries to find two target servers at a time among the currently active servers, which we call "a local bruteforce approach." One possible way of enumerating all possibilities of assigning two copies of a tenant to N servers will now be briefly discussed. Given that both copies of a tenant must be on different servers and that both copies are equivalent, all possibilities to assign

5.1 Algorithms for Static RTP

a single tenant to two servers can be done by creating a matrix with dimensions $N \times N$. The elements of this matrix are defined as follows:

$$(i,j) = \begin{cases} 1 \text{ if } i < j \\ 0 \text{ otherwise} \end{cases}, i,j \in N$$

An example for $N = 4$ servers:

$$\begin{pmatrix} 0 & 0 & 0 & 0 \\ 1 & 0 & 0 & 0 \\ 1 & 1 & 0 & 0 \\ 1 & 1 & 1 & 0 \end{pmatrix}$$

The number of possibilities for assigning two copies of a tenant to N servers is thus $N(N-1)/2$. This times the complexity of robustfit-static-mirror and robustfit-static-interleaved is complexity of robustfit-static-2atonce. In case no server pair can be found such that the two copies of the tenants can be placed in a valid way, two new servers are created. Similar to the previous algorithms, this algorithm places additional replicas individually for tenants requiring more than two replicas in a final step. robustfit-static-2atonce thus also naturally interleaves tenants but it has a higher computational complexity.

5.1.2 Meta-heuristics Based on Randomness

Having considered fast greedy heuristics, we consider a computationally more expensive heuristic next. We propose a variant of Tabu search [51], which we call tabu-static. This algorithm is a local search improvement heuristic. Given a starting solution (obtained by one of the greedy heuristics above), tabu-static search tries to remove an active server i by traversing the search space as follows. Every valid solution of RTP is a point in the search space. We move from one valid solution to another valid solution by moving a tenant t from i to a different server j, even if this move leads to an invalid placement. Next, we repair the placement (if possible without placing any new tenants on s). In order to avoid both cycling and stalling in a local optimum, a so-called Tabu list is used to store each move (t,i,j). We only allow a move if it is not already in the Tabu list. When the list reaches a certain length, the oldest element is removed to make room for the newest move (FIFO principle). The search aborts if—after a certain number of iterations—no placement was found in which all tenants on i could be moved to other servers. If a solution without server i was found, search continues from the new solution with the goal of removing another server.

The performance of the above algorithm relies on the careful adjustment of its parameters: the length of the Tabu list, the choice of server(s) to be cleared out, the order of tenants to be moved, the approach to fixing conflicts in invalid solutions, and the number of restarts. We experimentally set those parameters using a trial-and-error approach.

Another variant, called tabu-static-long, is parameterized to run longer and thus visit more solutions during the Tabu search. This variant is useful in scenarios where more migration budget is available.

5.1.3 Exact Algorithm

In the following, we propose a Mixed Integer (Linear) Programming (MIP) model for RTP that is closely related to the non-linear assignment formulation of RTP presented in Chap. 4. The choice of this type of formulation as opposed to others (e.g. column-generation based formulations [79]) was primarily motivated by its flexibility towards changing requirements. For example, it is rather simple to incorporate different optimization goals without complete remodeling. In order to model RTP as a MIP, we need to linearize the constraints in the assignment formulation of RTP.

In the following, we present linearizations of the non-linear constraints of RTP that were introduced in the previous chapter. Our MIP formulation can then be used with a variety of MIP solvers such as SCIP [3] or IBM ILOG CPLEX [59]. We focus on non-trivial linearizations in the following. In particular, we omit the linearization of logical constraints, for which the above-mentioned solvers provide built-in functionality.

We begin with Constraint (4.6), in which we set the value of p_i. Instead of one constraint with a maximum function, we need one constraint for each pair of server i and another server j. We thus arrive at $|N| - 1$ constraints of the following type:

$$\sum_{t \in T} \sum_{k \in R} \sum_{k' \in R} \frac{\ell(t)}{r(t)^2 - r(t)} \cdot y_{t,i}^{(k)} \cdot y_{t,j}^{(k')} \leq p_i$$

$$\forall i \in N, \forall j \in N, i \neq j \tag{5.1}$$

Because of the inequality, these constraints force p_i to be at least as large as in the original formulation in Constraint (5.1) that uses the maximum function. However, since the overall problem asks for a minimization of the number of active servers, even with this less restrictive constraint, p_i will assume the smallest possible value. Note that Constraint (5.1) is still non-linear due to the product of the two y variables. This product can be linearized by introducing an auxiliary variable $z \in \{0, 1\}^{T \times R \times R \times N \times N}$ as follows:

$$z_{t,i,j}^{(k,k')} = 1 \Leftrightarrow y_{t,i}^{(k)} \cdot y_{t,j}^{(k')} = 1 \quad \forall i, j \in N, \forall k, k' \in R, \forall t \in T \tag{5.2}$$

5.2 Algorithms for Incremental RTP

In order to ensure that placements are flexible in the sense of Definition 4.4, we introduced Constraint (4.13). For our MIP model, we linearize this constraint as follows.

$$\sum_{i \in N} \left(\sum_{k \in R} y_{t,i}^{(k)} \geq 1 \right.$$

$$\left. \wedge \sum_{t' \in T} \sum_{k \in R} \frac{\ell(t')}{r(t)} \cdot y_{t',i}^{(k)} + p_i \leq \mu \cdot \text{cap}_\ell(i) \cdot s_i \right) \geq 1$$

$$\forall t \in T \quad (5.3)$$

Constraint (5.3) contains multiple auxiliary conditions, such as $\sum_{k \in R} y_{t,i}^{(k)} \geq 1$. These auxiliary conditions are interpreted as 0/1 variables: they assume a value of one when the logical expression evaluates to true, and a value of zero otherwise. This pattern is called reification. Using reification, the linearization of the logical constraint is straight-forward. The existence quantor expression ensuring that there must be at least one server for every tenant in Constraint (4.13) can be replaced with a sum over reified constraints.

To obtain recoverable placements (cf. Definition 4.3), Constraint (4.12) can be linearized analogously.

Note that the size of the MIP formulation increases quickly in terms of the number of input parameters. For example, in order to linearize the product among the y variables in Constraint (5.1), we have to introduce $|N|^2 \cdot |R|^2 \cdot |T|$ binary auxiliary variables. Further extensions of the model would require even more auxiliary variables. In particular, computing the number of replicas per tenant $r(t)$ as part of the MIP (not done currently) would increase the size of the MIP beyond computational feasibility.

5.2 Algorithms for Incremental RTP

The static algorithms presented in the previous section are the foundation for our incremental placement algorithms. In order to leverage the different static algorithms, we introduce a generic framework used by all incremental placement strategies. The advantage of this framework is that it significantly reduces the solution search space, resulting in reasonable algorithm execution times.

5.2.1 A Framework for Incremental RTP

The framework consists of six phases. They are executed at the beginning of each reorganization interval, independent of the algorithm that is currently run. Individual algorithms must provide a method for placing a single replica of a tenant, which is "plugged" into the framework Such a plug-in method could, for example, be the robustfit-single-replica method described above. An incremental algorithm can also provide an own implementation for individual phases of the framework. The six phases of this framework are as follows.

(i) **Delete unnecessary replicas.** When the load of a tenant has decreased in comparison to the previous interval, it might be the case that removing a replica of the tenant is possible (see also the discussion on the lower bound on the number of replicas in Sect. 4.1.2). Therefore, in this phase, a heuristically selected replica of all tenants meeting this condition is deleted. Note that deleting a tenant does not count towards the migration budget.

(ii) **Ensure migration flexibility.** This phase ensures that all tenants have at least one replica on a server that has enough spare capacity to participate in a migration as a source server (Constraint (4.16)). For determining this server, the plugged-in algorithm is used. The goal is that the placement is flexible enough to migrate tenants without causing SLO violations.

(iii) **Create missing replicas.** This phase handles the opposite case of Phase (i), where the lower bound on the number of a tenant's replicas has increased as a result of higher load. The plugged-in algorithm is used to place enough additional replicas as necessary to match the new lower bound.

(iv) **Repair overloaded servers.** This phase repairs overloaded servers by migrating tenants away from them until they are no longer overloaded. The plugged-in algorithm is used to determine the target servers for replicas that are removed from overloaded servers.

(v) **Reduce number of active servers.** All servers are ordered by total load plus penalty. Then, all tenants on the most lightly loaded server are moved to other servers using the plugged-in algorithm. This phase is repeated with the next server up to the point where a server cannot be emptied without creating a new server.

(vi) **Minimize maximum load.** When it is no longer possible to reduce the number of servers, this phase flattens out the variance in load (plus penalty) across all servers. The goal is to avoid having servers in the placement that have a much higher penalty than other servers. Again, the plugged-in heuristic is used. This phase terminates when the migration budget is exhausted or further migrations would have too small an effect on the variance of load across servers.

The execution of the framework is immediately aborted when the migration budget is exhausted. When too low a value for the migration budget is chosen, the placement may be invalid after premature termination (i.e. it does not satisfy all of the constraints specified in Chap. 4). After completion of Phase (iv), however, a placement is always valid.

Note that the order in which the phases of the above framework are executed is in itself a heuristic. For example, experimentation has revealed that executing Phase (iv) after Phase (ii) results in fewer servers than the inverse order. The reason is that some overloaded servers are repaired as a side product of finding a safe migration source for tenants.

Note further that the question of deciding how many replicas a tenant should have is independent of this framework. Similar to algorithms for placing individual replicas, different strategies for determining the replication factor can be plugged in. The standard method is to use exactly as many replicas as suggested by the lower bound. Another method is to increase the lower bound by a fixed offset. A more sophisticated method would be to set the number of replicas across all tenants in a way that all replicas receive more or less the same load. A last method would be to repair overloaded servers in Phase (iv) by creating additional replicas elsewhere, thus decreasing the load of the tenant on the overloaded server.

In the following we discuss the plugin algorithms that we have developed for this framework.

5.2.2 Greedy Heuristics

The simplest (and also the fastest) algorithm is robustfit-inc. This algorithm merely comprises robustfit-single-replica, the method for placing a single replica (which has been described in Sect. 5.1.1). This method is plugged into the above framework as is.

Based on the observation that the space of possible actions when transforming a given placement into a new placement is very large, we created splitmerge-inc. This algorithm acts exactly as robustfit-inc. but, in terms of the framework above, provides an own implementation of Phase (iv), in which overloaded servers are repaired, and Phase (v), in which the framework tries to reduce the number of required servers. In Phase (iv) the only allowed operation is splitting each overloaded server into two servers. In Phase (v), conversely, merging two servers into one is the only legal operation, although multiple server pairs can be merged in one step. Since the underlying robustfit-single-replica method is very fast, we use a more complex procedure for deciding what servers to merge: splitmerge-inc. builds up its list of merge pairs by checking whether two servers can be merged for all candidate pairs in $N \times N$. The method in splitmerge-inc. for removing servers is effective, yet computationally intensive, since splitmerge-inc. actually tries to merge each pair of servers and selects best possible combination of merge pairs.

The approach towards repairing overloaded servers in splitmerge-inc. is rather simple. Essentially, overloaded servers are repaired by creating one new server per overloaded server. This might result in sharp increases in the number of servers between two reorganization intervals (i.e. up to twice the number of servers). Based on the intuition that this might be too drastic in many cases, we created a hybrid algorithm, called robustfit-merge. This algorithm replaces the implementation of Phase (iv), in which overloaded servers are repaired, with the standard one again and used robustfit-single-replica as the plug-in heuristic. robustfit-merge keeps the implementation of Phase (v) in splitmerge-inc. for reducing the number of servers by allowing only merges between servers.

5.2.3 Meta-heuristics Based on Randomness

We also re-use our Tabu search for the incremental version of RTP. We created the algorithm tabu-inc., which also uses robustfit-single-replica as its plugin heuristic. tabu-inc. replaces Phase (v), in which the number of servers is minimized, with the Tabu search described in Sect. 5.1.1. Note that tabu-inc. does not use a solution obtained by a greedy heuristic as the starting solution; it simply starts with the given placement. We simply omit Phase (vi), in which the variations in load are smoothed out until the migration budget is exhausted. This saves migration budget and thereby allows the Tabu search to run longer and visit more solutions. The next heuristic, tabu-inc.-long works exactly as tabu-inc., except that the parameters of the Tabu search are set in such a way that it runs significantly longer (and thus visits more solutions).

Finally, we propose another hybrid algorithm, called tabu-robustfit-inc. This algorithm combines robustfit-inc. with tabu-inc. . It runs robustfit-inc. as a preprocessing step, thereby omitting Phase (vi) so that the remaining migration budget can be used by tabu-inc. to improve the solution found by robustfit-inc.

tabu-robustfit-inc.-long is a variant of the tabu-robustfit-inc. algorithm that permits the Tabu component to run longer (similar to the static algorithm tabu-static-long).

5.2.4 Portfolio Approach

Portfolio approaches are meta-algorithms that combine as many algorithms as possible to arrive at the best possible solutions. Such algorithms have successfully been used for solving other $\mathcal{NP}$-complete problems. We thus created an own portfolio approach for RTP, called portfolio-inc., which combines all heuristics for incremental RTP. In each reorganization interval, portfolio-inc. runs all incremental heuristics and selects the best solution (i.e. the solution requiring the smallest number of servers). Note that choosing the best solution as the next solution is itself a heuristic approach. Instead of simply running all algorithms, more sophisticated techniques based on statistical machine-learning could be applied, to decide what algorithm should be run in a given reorganization interval. This method has been shown to work well for other optimization problems [90, 124], but not in scenarios comparable to incremental RTP. This is a potential avenue for future work.

5.2.5 Exact Algorithm

In the following, we present our MIP formulation for incremental RTP. Similar to our MIP formulation of static RTP, we restrict ourselves to discussing only the non-trivial linearizations of the constraints of incremental RTP. Our formulation relies on the introduction of an auxiliary variable $m \in \{0, 1\}^{T \times N}$. $m_{t,i}$ is set to one if and only if tenant t is assigned to server i in the new placement y but was not assigned to i in the previous placement y'. This is expressed in Eq. (5.4).

$$\left(\sum_{k \in R} y_{t,i}^{(k)} = 1 \wedge \sum_{k \in R} y_{t,i}'^{(k)} = 0 \right) \iff m_{t,i} = 1$$

$$\forall t \in T, \ \forall i \in N \quad (5.4)$$

The auxiliary variable $m_{t,i}$ is fundamental for expressing T_{mig}. Constraint (5.5) linearizes Constraint (4.15), which ensures that only a fixed volume of data δ is moved or copied when transitioning from y to y'. Arbitrary deletions of tenant replicas are allowed by this constraint.

$$\sum_{t \in T} \sum_{i \in N} \sigma(t) \cdot m_{t,i} \leq \delta \quad (5.5)$$

Constraint (5.6) linearizes Constraint (4.16). Again, we make use of reification. Whenever a tenant t appears on a server to which it had not been assigned in y' (indicated by $\sum_{i \in N} m_{t,i} = 1$), then there must be at least one server from which this tenant can safely be copied without SLO violations. We require that tenant t must have resided on such a safe migration source server in y'. Furthermore, the actual capacity of this migration *source* server drops to $\mu \cdot \text{cap}_\ell(i)$. Remember that this constraint might render the problem infeasible (see discussion in Sect. 4.2).

$$\sum_{i \in N} m_{t,i} \geq 1 \Rightarrow$$

$$\left(\sum_{i \in N} \left(\sum_{k \in R} y_{t,i}'^{(k)} = 1 \right. \right.$$

$$\left. \left. \wedge \sum_{t' \in T} \sum_{k \in R} \frac{\ell(t')}{r} \cdot y_{t',i}'^{(k)} + p_i \leq \mu \cdot \text{cap}_\ell(i) \right) \geq 1 \right)$$

$$\forall t \in T \quad (5.6)$$

Constraint (5.7) expresses the temporary capacity drop during the process of migrating tenants onto a *destination* server of a migration. Whenever a tenant t appears on a server to which it had not been assigned in y' (indicated by $\sum_{i \in N} m_{t,i} = 1$), the capacity of this server is reduced to $\nu \cdot \text{cap}_\ell(i)$. Again, this is because the server must deal with the incoming tenant migration and can consequently handle fewer requests. Note that—with a practically relevant number of tenants and servers—the MIP model becomes so large that it is computationally challenging to solve instances of incremental RTP.

$$\sum_{t \in T} m_{t,i} \geq 1 \Rightarrow$$

$$\sum_{t \in T} \sum_{k \in R} \frac{\ell(t)}{r} \cdot y_{t,i}^{(k)} + p_i \leq v \cdot \mathrm{cap}_\ell(i) \cdot s_i$$

$$\forall i \in N, \quad m_{t,i} \in \{0, 1\} \tag{5.7}$$

Having described our algorithms in this chapter, we will provide a detailed experimental evaluation of the algorithms in the next chapter.

Chapter 6
Experimental Evaluation

In this chapter, we evaluate our algorithms for RTP. This is done by replaying the load changes of all tenants over a 4-months-period against our algorithms. The load changes are aggregated in 10 min intervals. For each 10 min interval, a new instance of RTP is created and our different algorithms produce a series of migrations which must be completed within a 10 min window. Since our evaluation is based on the real-world load traces presented in Sect. 2.4, we begin with describing how we bootstrapped these traces to generate a sufficiently large number of tenants to test our algorithms at scale, while ensuring that the resulting set of tenants is realistic (Sect. 6.1). Afterwards, we present a detailed series of experiments characterizing the performance of our various algorithms in terms of solution quality and speed (Sect. 6.2).

6.1 Tenant Trace Data Used for Experimentation

As mentioned in Sect. 2.4, we obtained a sample of 87 randomly selected tenants in Europe and the U.S. for the SAP application. We were given only a subset of the data to protect the customers' privacy. While this sample is representative, it is not large enough to test our placement algorithms at scale. We obtained the anonymized application server log records for a 4-months-period as well as additional statistics about the tenants (e.g. their database size). These statistics suffice to compute the load of each tenant following the methodology described in Chap. 3. In order to allow meaningful experiments, we used a methodology called *bootstrapping* to create new tenant traces from the ones we obtained from SAP. We adapted the bootstrapping process by Bodík et al. [16] such that:

1. We arrive at a sufficiently large number of tenants;
2. The load of each tenant follows either the temporal characteristics of a working week or the characteristics of a demo system (cf. Sect. 2.4);

3. Each tenant has a realistic size; and,
4. Variations in load differ sufficiently among tenants.

The latter requirement is challenging. On the one hand, our goal is to create new tenants with a realistic load pattern. This requires that the pattern resembles the original tenants. On the other hand, too much resemblance among tenants is not desirable, since as a result the aggregate load in the cluster would be smoothened (cf. Fig. 1.1) as the total number of tenants is increased. Our bootstrapping technique takes this trade-off into account, as we shall see in the following.

6.1.1 *Bootstrapping Process*

To generate new tenants, we adopt the following methodology. Each original (parent) tenant trace is bootstrapped into c child tenants according to the trace bootstrap process below, where c is the number of bootstrap copies (a tunable parameter). Each child tenant is then given a size drawn at random from a list of all parent tenant sizes. In our bootstrapping, we differentiate between regular tenants and demo tenants and do one bootstrapping process for each type of tenant. Regular tenants all have comparably similar request rate patterns but highly varying sizes. Demo tenants have irregular usage patterns but exhibit smaller difference in their sizes (i.e. they are generally small). Therefore, it seems reasonable to choose size independently of pattern sizes.

The process of bootstrapping a parent trace into a single child trace, shown in Fig. 6.1, is as follows:

1. The parent trace is divided into equally sized subintervals of size W. The last subinterval of the trace may contain less than W entries. In the following, we will refer to such subintervals as *window*.
2. The indices of a window of the parent trace correspond directly to the indices of a window in the child trace. For example, the first window spans $\{0, 1, \ldots, W-1\}$, the second $\{W, W+1, \ldots, 2W-1\}$, etc., in both parent and child traces.
3. For each parent window, W values are chosen at random with replacement and uniform probability. These values are placed into the corresponding child window in the order that they were chosen.

The quality of the result of the bootstrapping processing depends on the choice of W. We will describe how to obtain a good value for W in Sect. 6.1.3.

This process results in c times many new tenants, each with a realistic size and a trace that is similar but not identical to its parent. In our case, the bootstrapping process generates $87 \times c$ many new tenants.

6.1 Tenant Trace Data Used for Experimentation

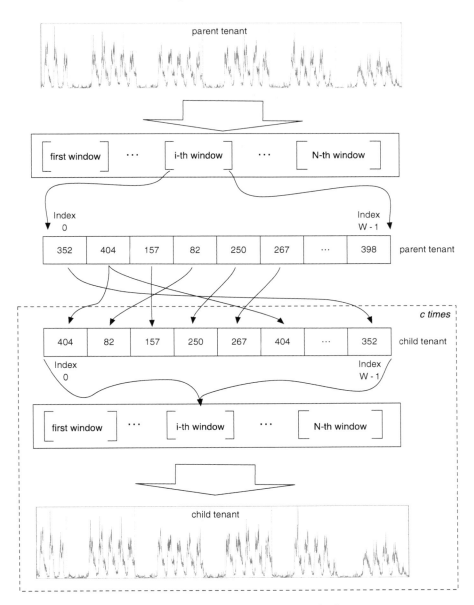

Fig. 6.1 The bootstrapping process for generating realistic tenant load traces

6.1.2 Tenant Sign-Ups

In a real-world system, new tenants join the system throughout the course of its life. Unfortunately, we only received traces of tenants that were present at the start of the trace recording. Therefore, we decided to simulate the growth of the customer base by adding new (bootstrapped) tenants at random points in time, growing the number of tenants in the system at a rate of 1–2 % per week, which is the sign-up rate of the SAP application for which we obtained the log data.

6.1.3 Choosing the Bootstrap Window Size

The balance of realism and similarity to the parent in the child trace is determined by the bootstrap window size W. With a very small W, we run the risk of all or most $c + 1$ parent and child traces of a single tenant trending in unison, as a load spike in the parent may occur in its children around the same 10 min intervals. Conversely, a large W causes irregularity and lack of natural smoothness in the children: a window may fall on the boundary between night and workday, and the low-request night 10 min intervals and high-request workday 10 min intervals become intermingled, giving the trace an unnatural "spikiness." The goal is thus to set W such that the deviation between parent and children is high enough while no unnatural spikiness occurs. In the following we analyze both phenomena.

Figure 6.2a shows how the similarity between parent and child traces decreases as the size of the bootstrap window increases. Specifically, it shows the mean deviation in load across all 10 min intervals between the parent and its bootstrapped children, computed using the root mean square deviation (RMSD) for each parent/child pair. This is shown for one of the largest tenants in our trace, which frequently exceeds the capacity of a single server. The increase of the RMSD flattens out beyond $W = 10$.

Figure 6.2b characterizes unnatural spikiness in the bootstrapped children. When looking at the amount of change that a parent exhibits between two 10 min intervals (i.e. the *delta* in request rate), we define a delta as unnatural when it is larger than the 75-th percentile of all deltas of the parent. Figure 6.2b, again focusing on a single tenant, shows the percentage of deltas larger than the 75-th percentile for the parent and its children with varying W. For the parent, obviously, this percentage is independent of the bootstrap window size. For the children, the percentage of unnaturally high deltas is the same as for the parent when choosing $W = 1$. For $2 \leq W \leq 10$ it is lower than for the parent. This is because with such small window sizes there is always a relatively high chance for picking a value that has already been picked previously in the same window as the next value. For larger values of W, the children have higher deltas than the parent, i.e. the children become more unnatural. This is because for large window sizes the chance increases that a vastly different value from the value picked for the current 10 min interval is contained in the window and is picked as the value for the next 10 min interval at some point.

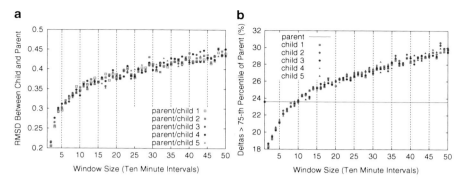

Fig. 6.2 Choosing the optimal bootstrap window size W. (**a**) RMSD between parent and child increases with window size: the child is less similar to the parent. (**b**) The difference in requests between ticks of children, compared to the 75-th percentile of the parent

Based on this analysis, we chose a bootstrapping window size of $W = 10$. This choice balances deviation of a parent and its children with unnatural changes between two 10 min intervals.

6.2 Evaluation of Algorithms for RTP

In the following, we evaluate our algorithms for RTP, focusing mainly on the incremental variant. The evaluation is structured as follows: Sect. 6.2.1 discusses the performance of our algorithms with respect to their ability to balance (i) the number of active servers, (ii) computation time, and (iii) their abilities to react to load changes. We will see that robustfit-inc. achieves a good balance between all three measures. Consequently, Sect. 6.2.2 explores robustfit-inc. further. We investigate lower bounds for server cost. We also study the consequences of increasing the number of replicas per tenant beyond the minimum, which has some interesting effects on the number of active servers and the stability of a placement over the day. In Sect. 6.2.3 we develop generic extensions for our heuristics to reduce the impact of temporarily overloaded servers until it becomes negligible. These extensions are orthogonal to our placement algorithms. We evaluate these strategies with robustfit-inc. Afterwards, we show that these extensions are also useful when dealing with an increasing number of simultaneous server failures.

Unless otherwise noted, we use a dataset with $c = 5$ bootstrap copies per original tenant, which results in a total number of 522 tenants. In all experiments, we assume that servers have a DRAM capacity of $cap_\sigma = 32\,GB$. We further assume a load capacity of $cap_\ell = 1.0$ per server. We further assume that all servers are homogeneous in terms of their capacities. While our algorithms also work for heterogeneous servers, this assumption considerably simplifies the exposition of our experiments. Load changes are observed in intervals of 10 min, since shorter intervals could result in "thrashing" in the sense of overreacting to short-lived load bursts. We thus set the migration budget to $\delta = 27$ GB: based on our analysis of migration durations in Sect. 3.4.2, this amount of data can safely be migrated in a 10 min interval using SAP's in-memory database and a 10 Gb/s Ethernet interconnect.[1]

Most of the experiments shown in the following were conducted by replaying a typical working day, a Wednesday, against our algorithms. We also ran all experiments across the full 4-months-period as well and observed the same relative performance among our algorithms than for the chosen Wednesday. Again, reporting results for a single day considerably improves the exposition of our results.

Our experiments were conducted on an Intel Xeon X7560 server with 2.27 GHz running Linux. We implemented our heuristics in the Scala language [89] and used CPLEX [59] as a MIP solver. We have not yet parallelized our algorithms. We would expect a significant speed-up from a multi-threaded implementation, especially for Tabu search and portfolio-inc. We consider this part of future work.

6.2.1 Comparison of Heuristics for RTP

In order to evaluate our heuristics for solving (incremental) RTP, we consider the following three measures:

1. The *cost* associated with the resulting placements, measured as accrued when using a varying number of "high memory" instances on Amazon EC2 [5];
2. The *computation times* required by an algorithm to produce a placement; and,
3. The *robustness* of the resulting placements towards abrupt load spikes.

Not all can be optimized for at the same time; consequently, a trade-off between these measures must be found. A particularly low cost placement may require an unrealistic amount of computing time and then, at the same time, the tenants might be packed so tightly that servers are prone to temporary overloads when load increases.

Experiment (i):Incremental Placement Vs. Static Provisioning for Peak Load

Table 6.1 summarizes the benefits of our incremental algorithms over the conventional static approach measured on a typical working day. The conventional approach, modeled after Yang et al. [125], entails monitoring all tenants for 1 week and observing the peak load of each tenant within that period. Afterwards, one provisions for this peak load. We used the week directly preceding the Wednesday chosen for our experiments to estimate the maximum load for each tenant. We solve static RTP for the observed peak loads using greedy heuristics to determine a layout.

---

[1] We study the effect of changing δ in Experiment (iv).

6.2 Evaluation of Algorithms for RTP

Table 6.1 Server cost and running time of heuristics for RTP

Algorithm	Cost	Servers Max	Running time Avg	Max
Static:				
robustfit-static-mirror	$3,456.00	320		66.4 s
robustfit-static-interleaved	$2,073.60	192		481.3 s
Incremental:				
tabu-inc.	$273.83	40	2.5 s	5.9 s
tabu-inc.-long	$208.20	34	26.8 s	87.0 s
tabu-robustfit-inc.	$202.95	33	3.0 s	10.8 s
robustfit-inc.	$201.45	39	1.7 s	3.7 s
splitmerge-inc.	$200.18	38	95.5 s	321.6 s
robustfit-merge	$198.08	32	84.2 s	256.4 s
tabu-robustfit-inc.-long	$193.05	33	19.8 s	60.5 s
portfolio-inc.	$191.55	33	182.1 s	565.3 s

Table 6.1 shows that robustfit-static-mirror, the simplest static algorithm, requires 320 servers, whereas robustfit-static-interleaved, which interleaves tenant replicas, uses 192 servers. These static placements have a daily operational cost of $3,456 (or $2,073, respectively).

In contrast, our incremental algorithms, which alter the placement in 10 min intervals and use the lowest possible number of replicas per tenant in each 10 min interval, require between 33 and 40 servers at most during times of peak load and much fewer servers during the night and times of low load (e.g. on weekends). Table 6.1 shows the cost for server rent for all incremental algorithms. On average, cost is an order of magnitude lower when using an incremental algorithm as opposed to static provisioning based on peak load. The cost for server rent on our exemplary Wednesday is a factor of 18.0 lower for robustfit-inc. in contrast to the static placement obtained with robustfit-static-mirror. Similarly, robustfit-inc. achieves a factor of 10.3 cost improvement over the static algorithm robustfit-static-interleaved (our best static algorithm). Note that these factors should be viewed as an approximate baseline for the potential cost savings of our approach; the reason is that they depend on (i) the business day we chose for our experimental evaluation and (ii) the days we chose for monitoring the tenants in order to observe their peak load requirements. Since our experimental data is based on the log data described in Sect. 2.4, we know that the maximum aggregate load of all tenants is similar for all business days (cf. Fig. 1.1). To exclude the possibility that this result is biased because (i) tenant load on our chosen Wednesday is low in comparison to the working days of the previous week or (ii) the previous week contains a working day where tenant load is particularly high, we repeated the experiment with the following variations. We benchmarked robustfit-inc. against static placements with peak load requirements obtained from analyzing (i) the same day on which we run our incremental algorithms (Wednesday) and (ii) the day before (Tuesday). Also for these cases, the factor by which robustfit-inc. improves costs in comparison to the static algorithms varies between 6.8 and 11.6.

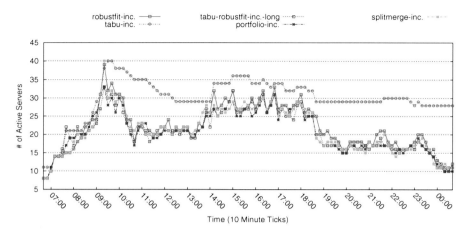

Fig. 6.3 Number of active servers on an average day for selected algorithms

Cost also varies substantially among the different incremental algorithms. Specifically, cost varies from $191.55 to $273.83, a range of 30 %. Within that range, the algorithms can be grouped into clusters around the cost ranges of $191–$193 and $198–$203. In terms of cost, the randomized algorithms tabu-inc. and tabu-inc.-long do worst, with tabu-inc. performing particularly poor. Experiment (ii) discusses the differences among the incremental algorithms in greater detail.

Experiment (ii):Incremental Algorithms: Cost vs. Running Time

In this experiment, we benchmark the running times of our incremental heuristics against the associated costs of the resulting placements. To better visualize the quality of the placements over time, we also plotted the number of active servers for selected incremental algorithms throughout the day in Fig. 6.3. The night time is omitted from the chart, since all incremental algorithms require the same number of servers at night in most cases. The time in a 10 min interval is conceptually split into the time for algorithmic computation and the remainder, which is used to physically carry out the migrations. The shorter the running time of an algorithm the more time is available for performing migrations. A short running time also indicates good scalability of an algorithm towards larger problem instances.

Among the fast algorithms with an average running time below 10 s (see Table 6.1), robustfit-inc. finds the placements with the lowest cost. It is also the fastest algorithm overall, and thus the best option for short reorganization intervals.

Among the longer-running heuristics, portfolio-inc. naturally delivers the best results because it combines all other incremental heuristics and selects the placement with the fewest number of server in each 10 min interval. Consequently, portfolio-inc. is also by far the slowest algorithm. tabu-robustfit-inc.-long is almost as good as portfolio-inc. with regard to server cost. However, on average, tabu-robustfit-inc.-long is more than ten times faster than portfolio-inc. It is by far the fastest heuristic among those that produce placements with a daily server cost of around $190. This is because tabu-robustfit-inc.-long spends much more time in the phases involving random traversal of the search space than tabu-inc. and tabu-robustfit-inc. This additional time is sufficient to produce improvements of the solutions obtained by robustfit-inc. From a running time perspective, tabu-inc.-long is comparable to tabu-robustfit-inc.-long, however, the results produced by tabu-robustfit-inc.-long are much better. tabu-robustfit-inc.-long is the best choice if one can allow investing up to 1 min of computation per 10 min interval.

Figure 6.3 shows the number of required servers for selected incremental algorithms throughout the day. At certain times during the day portfolio-inc. produces placements requiring more servers than some of the other incremental heuristics (e.g. robustfit-merge). This phenomenon—counter-intuitive at first, since portfolio-inc. is supposedly the best incremental heuristic—highlights the strong influence that the given placement from the previous interval has on the ability of any incremental algorithm to minimize the number of active servers. This sensitivity results in portfolio-inc. requiring a higher maximum number of servers than robustfit-merge (see Table 6.1). splitmerge-inc., also shown in Fig. 6.3, produces placements of similar cost than robustfit-inc., although it requires a considerably higher amount of computation time; in fact is the slowest heuristic except for portfolio-inc. However, splitmerge-inc. has an important strength, as we shall see in the following experiments.

Experiment (iii): Robustness Towards Load Spikes

A crucial measure for comparing incremental heuristics is the robustness of the placements towards sudden increases in tenant load. In the following, we analyze the number of *temporarily overloaded* servers, i.e. servers that are overloaded directly after load changes have been observed but before the placement algorithm has run and tried to alleviate the overload situation.

When using an incremental placement strategy, one tries to find a placement using the minimal number of servers while still providing just enough resources to handle the load of all tenants without violating performance SLOs. This naturally results in situations where servers have few spare capacity. When changes in the load of the tenants are observed, a new incremental placement is computed and tenants are quickly migrated away from possibly overloaded servers. When using a static placement strategy, in contrast, all servers must have enough spare capacity to handle an estimated peak load across a longer time period. In this experiment, we study how many servers are temporarily overloaded in each 10 min interval when an incremental placement strategy is used. Here, a temporarily overloaded server is a server that is beyond its load capacity limit at the beginning of a 10 min interval, i.e. after new values for the load of the tenants have been observed and before a new incremental placement is computed and put in place. This metric is an indicator for the robustness of a placement towards unexpected load spikes. The fact that servers become temporarily overloaded while the placement is being reorganized in response to a load spike is the most significant downside of incremental placement.

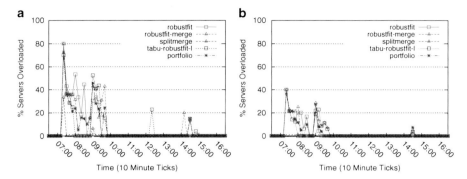

Fig. 6.4 Percentage of overloaded servers. (**a**) Including penalty. (**b**) Without penalty

Managing the trade-off between temporary overloads and cost for server rent is a key challenge (which we will address in Sect. 6.2.3).

Figure 6.4 shows what fraction of all servers is temporarily overloaded across the 10 min intervals of our exemplary business day. We focus on a few selected incremental heuristics. Figure 6.4a includes failure penalty (cf. Eq. (4.14)) in the calculation of overloaded servers. Figure 6.4b, in contrast, only takes the actual load on the servers into account. We observe that temporarily overloaded servers occur mostly due to drastic increases in load during peak hours, e.g. early in the morning and after lunch. The graphs focus on those times of the day and omit the evening and night hours, where no temporary overloads occur. In the morning at around 07:20 a.m. such a spike occurs and the heuristics have to react by expanding the cluster. At this point in time, almost all algorithms find 70–80 % of all servers temporarily overloaded when considering penalty, or up to 40 % without penalty, respectively. Placements computed with the splitmerge-inc. algorithm seem very robust towards load spikes: only 30 % of the servers are beyond their load capacity when considering penalty, and no server at all is overloaded when considering only the actual load without a possible server failure. Surprisingly, when not considering penalty, there is not a single 10 min interval with a temporarily overloaded server when using splitmerge-inc. While one would expect that the percentage of overloaded servers decreases more quickly for splitmerge-inc. than for the other heuristics given the aggressive way in which splitmerge-inc. deals with overloaded servers, it is astonishing that the percentage of overloaded servers is also much lower initially than for the other heuristics. The reason is that splitmerge-inc. only removes a server if it can be merged with another server. When load decreases, the rate at which splitmerge-inc. removes active servers is thus slightly slower than for the other algorithms. In order for splitmerge-inc. to merge two servers, both servers must have a sizable amount of spare capacity. However, when two servers cannot be merged, there may still be enough spare capacity to provide a high robustness towards sudden increases in tenant load.

6.2 Evaluation of Algorithms for RTP

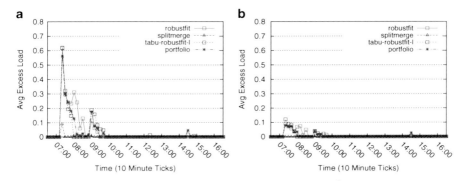

Fig. 6.5 Average excess load across servers. (**a**) Including penalty. (**b**) Without penalty

While the overall tenant activity (and thus overall load) continues to increase until around 10:00 a.m. (see also Fig. 6.3), the percentage of overloaded servers quickly decreases after its 07:20 a.m. spike for most algorithms. Between 07:20 and 09:00 a.m., robustfit-inc. produces placements with a higher percentage of temporarily overloaded servers than the other heuristics. Between 09:00 and 10:00 a.m. tenant load increases more steeply again and most algorithms produce placements with 30–50 % overloaded servers (including penalty), except for splitmerge-inc., which gets away with less than 10 %.

Given the considerable amount of servers which are temporarily beyond their capacity limit (e.g. and thus temporarily not adhering to the response time guarantee), we need to investigate how severe the overload situation is on the affected servers. Figure 6.5 shows the average excess load across all servers over time for a subset of our heuristics. Remember that the load capacity limit of a server in our experiments is $cap_\ell = 1.0$. Therefore, a value of e.g. 0.1 means that servers are overloaded by 10 % on average. When including penalty, servers are overloaded by around 60 % when the 07:20 a.m. load spike occurs (Fig. 6.5a) for most algorithms. Again, splitmerge-inc. exhibits a much higher robustness with the servers overloaded by less than 10 %. For the 09:00 a.m. spike, the average overload is between 10 (robustfit-inc.) and 20 % (others). For splitmerge-inc., the excess load is negligible. Note that while these numbers are relatively high, the inclusion of penalty assumes that one of the servers in the cluster has failed. In a failover scenario these numbers can be considered moderate. Figure 6.5b shows the average excess load as it occurs during normal operations. On average, no server is overloaded by more than 10 % for most algorithms (again, no overloads at all occur when using splitmerge-inc.).

Since one novel aspect of RTP is to provide response time guarantees even in spite of server failures, we investigate the 07:20 a.m. spike in more detail. Having established that 70–80 % of all servers are temporarily overloaded by approximately 60 % when including penalty, we are interested in understanding how this excess load is distributed across all servers in the cluster. Figure 6.6a shows the excess

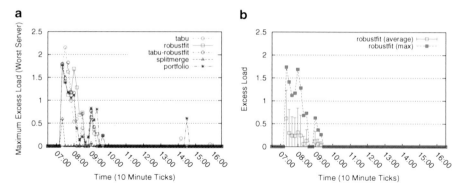

Fig. 6.6 Excess load on worst server. (**a**) Including penalty. (**b**) Including penalty, showing maximum, average and standard deviation for robustfit-inc.

load for the "worst" server in each 10 min interval. By worst server we mean the server with the highest amount of excess load among all servers in a placement in a given 10 min interval. Note that the worst servers for two adjacent 10 min intervals are not necessarily the same server. We use this metric to show the behavior of the worst case over time. During the 07:20 a.m. spike, most algorithms produce placements with a worst server having an excess load of around 1.7 (splitmerge-inc. again does much better with a value of 0.6). Figure 6.6b focuses solely on robustfit-inc. It shows the average excess load across time including the standard deviation across all servers (using error bars). It also shows the worst server for robustfit-inc. over time (maximum value for excess load). Given (i) the comparatively large gaps between the average excess load and the maximum (i.e. the worst server) and (ii) the relatively high standard deviations from the average, we conclude that the overall excess load in a 10 min interval is not evenly distributed across servers. The excess load on the worst server is thus a good indicator for the robustness of a placement towards sudden increases in load.

In summary, the performance of most algorithms is similar regarding temporarily overloaded servers, except for splitmerge-inc., which is clearly superior to all other heuristics. However, the high running times of splitmerge-inc. (between 1.5 and 5.5 min per 10 min interval) make its use mostly impracticable. Note also that splitmerge-inc. is harder to parallelize than for example tabu-inc. or portfolio-inc. due to its complex merge phase. Based on this experiment, it becomes clear that our heuristics must be extended to minimize the impact of temporarily overloaded servers. We develop and evaluate appropriate techniques in Sect. 6.2.3.

Experiment (iv): Varying the Migration Budget

All previous experiments have been conducted with a migration budget of 27 GB, which is motivated by practice. We consider 27 GB conservative, since it is based on our assumption that all migrations must be performed sequentially. However, in case an algorithm decides that tenants on more than two (or more) servers shall

6.2 Evaluation of Algorithms for RTP

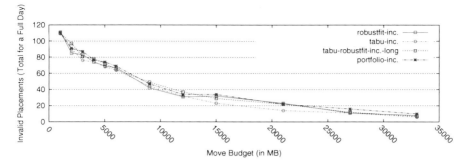

Fig. 6.7 Number of invalid placements with varying migration budget for robustfit-inc., tabu-inc., and tabu-robustfit-inc.-long

be migrated, there is an increasing potential that migrations can be performed in parallel. Scheduling multiple migrations to the maximum parallel extent under time constraints is itself an $\mathcal{NP}$-complete problem, which is beyond the scope of this dissertation. Anderson et al. [9] provide an excellent overview of the *data migration problem* and propose a polynomial time algorithm.

In this experiment, we investigate the case where shorter reorganization intervals become necessary and fewer data can be migrated during each interval in turn. As we shall see in Experiment (v), our algorithms for incremental RTP produce placements that require a similar number of servers as our static algorithms, which have an "unlimited" migration budget. Thus, we expect no substantial improvements from increasing the migration budget and focus on the effects of decreasing the migration budget in the following.

We focus on a single metric, the number of invalid placements across the day, i.e. the total number of 10 min intervals where the migration budget did not suffice for performing enough migrations so that all constraints of RTP are satisfied (cf. Definition 4.1 and Sect. 4.1). Figure 6.7 shows the total number of invalid placements for selected incremental algorithms. Given a large migration budget of 33 GB this number is around ten for all tested algorithms. Invalid placements occur mainly in 10 min intervals where load increases sharply, i.e. between 7:00 and 10:00 a.m., analogously to temporarily overloaded servers (cf. Experiment (ii)). With decreasing migration budget, the number of invalid placements naturally increases, since fewer changes can be made in each 10 min interval to adjust to load changes. All heuristics show a similar behavior when dealing with a decreasing migration budget. portfolio-inc. produces slightly more invalid placements than the other algorithms, since it always selects the solution with the fewest number of active servers in each 10 min interval and thus packs tenants tighter than all other algorithms. On average, tabu-inc. produces slightly fewer invalid placements than the other algorithms because it does not pack tenants as tightly as the other algorithms. tabu-inc. should thus be favored over robustfit-inc. when given a small migration budget.

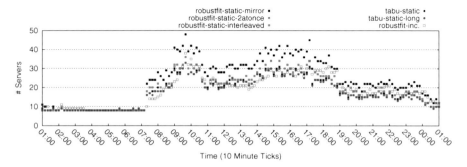

Fig. 6.8 Number of servers required by algorithms for static RTP in comparison to robustfit-inc.

We conclude that robustfit-inc. provides the best balance between server cost, running time, and robustness towards temporary load spikes. The most robust algorithm, splitmerge-inc., has prohibitively long running times. Therefore, we evaluate robustfit-inc. in more detail in the following.

6.2.2 Advanced Experiments with Robustfit-Incremental

Having established that immense cost savings can be realized when using incremental placement techniques, the obvious next question is by how many servers the placements obtained by our heuristics deviate from placements which are *optimal*. We therefore investigate lower bounds on server cost in this section. We also consider varying the number of replicas per tenant beyond the minimum. Since robustfit-inc. provides a good trade-off between cost, computation time, and robustness, we restrict our focus to robustfit-inc. going forward.

Experiment (v):Lower Bounds on Operational Cost: Static RTP

In this experiment, we compare our heuristics for incremental RTP with a baseline computed by heuristics for static RTP. When comparing heuristics for incremental and static RTP, we assume that it is feasible to run a static algorithm in each 10 min interval. When doing so, we expect the static algorithms to achieve better results than the incremental ones, in contrast to Experiment (i), where the static heuristics have been run only once. We remind the reader that static RTP contains no notion of a migration budget or an existing placement to start from. Static algorithms thus cannot be applied when reorganization intervals are short. However, we are interested in analyzing how close our incremental algorithms get to the number of servers obtained by the static algorithms.

Figure 6.8 shows the number of servers over the day for all our static algorithms. The figure also shows the number of servers obtained by robustfit-inc. as a point of comparison.

6.2 Evaluation of Algorithms for RTP

On average, robustfit-inc. performs as good as the four best static algorithms (i.e. robustfit-static-2atonce, robustfit-static-interleaved, tabu-static, and tabu-static-long). This is surprising, given that—according to intuition—the incremental placement problem seems to be more challenging than the static one, given the additional constraints of a migration budget and decreasing server capacities during migrations. For some 10 min intervals, robustfit-inc. even requires fewer servers than the static algorithms (e.g. 7:00–9:00 a.m. and 12:00–14:00 p.m.); however, there are also cases where robustfit-inc. requires more servers (e.g. 5:00–7:00 p.m.). One reason for the good performance of robustfit-inc. might be that it often has the opportunity to start from a good solution obtained in the previous 10 min interval. Incremental improvements of good solutions are carried forward by robustfit-inc. The main insight from this experiment is that in terms of the number of required servers, there is no significant difference between static and incremental heuristics, although incremental RTP is a much more constrained problem.

Experiment (vi): Lower Bounds on Operational Cost: MIP Approach

Heuristics for static RTP provide a lower bound on the required number of servers that can be computed in a time span proportional to the problem size. It may, however, be possible to find valid solutions with fewer servers when exploring all combinatorial options of assigning tenants to servers in a systematic way. In the following, we do just that using our CPLEX implementation of the MIP formulation of RTP that was presented in Sects. 5.1.3 and 5.2.5. We are interested in studying the relative gap between the solutions obtained by CPLEX and the solutions found by our heuristics. Unfortunately, the standard problem size used in our experiments (i.e. five bootstrap copies per tenant) is too large, since the resulting MIP model contains so many variables that solving the model is computationally infeasible. We therefore use a smaller data set (with one bootstrap copy per tenant), on which we run both CPLEX and our heuristics for comparison. In total, the small data set consists of 174 tenants.

We have created a CPLEX implementation for both the static and incremental version of RTP, named *cplex-static* and *cplex-inc.*, respectively. Similar to the previous experiment, we assume that the static variant can be run for each 10 min interval, which is impracticable, as stated above. Yet, the results obtained by cplex-static provide a good lower bound for the number of servers that our algorithms could potentially achieve.

The efficiency of CPLEX can be drastically improved if it does not have to start from scratch when searching for a solution. The better a starting solution it is given, the faster its search through the combinatorial space. cplex-static begins with computing a solution using robustfit-static-interleaved, which is passed to the CPLEX solver. Similarly, in cplex-inc., we first run portfolio-inc. and pass the result (i.e. the best solution found after running all incremental algorithms) to CPLEX.

Figure 6.9a compares running cplex-static ("best integer solution") in each 10 min interval across our exemplary Wednesday to robustfit-inc. The number of servers obtained by robustfit-static-interleaved (the "starting solution") is also shown, as well as the best known lower bound obtained by CPLEX. It is guaranteed

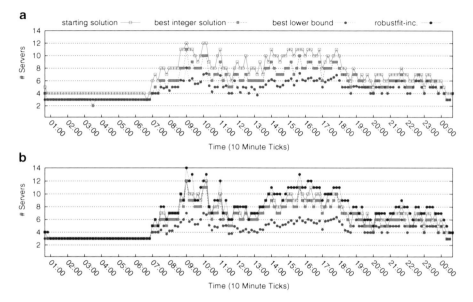

Fig. 6.9 MIP results for a small data set. (**a**) cplex-static. (**b**) cplex-inc. and robustfit-inc.

that no solution with fewer servers than the best lower bound exists. However, there is no guarantee that a solution with as few servers as the best lower bound does in fact exist. For 10 min intervals where the best integer solution and the best lower bound have the same value, we know that an *optimal* solution has been found. As can be seen in Fig. 6.9a, this is only the case for 10 min intervals with low overall load, i.e. at night and early in the morning. During times of high load, we observe a 30 % difference on average (42 % at most) between the best integer solutions and the best lower bounds. As mentioned above, however, this does not indicate that the best integer solutions are inoptimal or could be improved by 30 %. In fact, it is known that MIP formulations produce weak lower bounds for the related bin-packing problem, i.e. the lower bounds are often relatively far below the actual optimum [117]. Table 6.2 summarizes the quality of the solutions obtained by cplex-static in comparison to the starting solutions that it was given across the whole day. The best integer solutions require 1.5 servers fewer servers on average (and three servers fewer at most). Given the low total number of servers, which is due to the smaller data set used in this experiment, this difference is high when expressed as a percentage: the average improvement is 20.8 %, while the maximum improvement is as high as 50 %. The latter occurs at 03:30 a.m. (cf. Fig. 6.9a). This improvement comes at a significant price: we parameterized CPLEX in a way that allows for up to 24 h of computation (per 10 min interval) before the exploration of the combinatorial search space is aborted.

Figure 6.9b shows the same experiment for cplex-inc. Here, CPLEX adheres to the migration budget in all 10 min intervals and must deal with reductions in

6.2 Evaluation of Algorithms for RTP

Table 6.2 Improvements obtained by cplex-static and cplex-inc.

	Absolute		Relative	
	Avg	Max	Avg (%)	Max (%)
cplex-static:				
Improvement of best integer solution over starting solution	1.5 servers	3 servers	20.8	50.0
cplex-inc.:				
Improvement of best integer solution over starting solution	0.6 servers	2 servers	8.4	28.5
Improvement of best integer solution over robustfit-inc.	1.3 servers	4 servers	15.0	42.9

server capacities during migrations. The starting solutions are computed using our portfolio-inc. algorithm. Since we were not able to compute the solutions for multiple 10 min intervals in parallel (which can easily be done for cplex-static, which is not dependent on the result for the previous interval), we allowed only 3 h of computation before CPLEX aborts. As shown in Table 6.2, cplex-inc. improves the starting solutions by 0.63 servers on average across all 10 min intervals, which is an average improvement of 8.4 %. The maximum amount by which cplex-inc. improves the starting solutions is two servers. Given the low total number of servers, the maximum improvement is as high as 28.5 % in relative terms. The relative difference in these metrics between cplex-inc. and robustfit-inc. is approximately twice as high as between cplex-inc. and the starting solution. Note that the gap between portfolio-inc. and robustfit-inc. is much less pronounced in the experiments that we conducted on the larger data sets.

In summary, the improvement obtained by cplex-static over the starting solution is 20.8 % on average, while cplex-inc. can only to improve the starting solutions by 8.4 % on average. This is an indicator that incremental RTP is indeed harder to solve than static RTP because of its additional constraints (i.e. the migration budget and reduced server capacities during migration). We hypothesize that the relative gaps between the best integer solutions of the MIPs and the solutions obtained with our heuristics are similar for larger problem sizes (such as our larger data set with five bootstrap copies per tenant). This hypothesis is motivated by the related bin-packing problem, for which these relative gaps are actually constant as problem size increases [48].

We conclude that our heuristics for both static and incremental RTP find placements of high quality in comparison to baseline results that require an excessive amount of computation.

Experiment (vii): Varying the Number of Tenant Replicas

In all previous experiments, the number of replicas per tenant was set to the minimum number of replicas required such that each replica fits on a single server. This lower bound on the number of replicas per tenant was discussed in Sect. 4.1.2. In Sect. 5.2.1, we listed several meta-heuristics for dynamically computing the

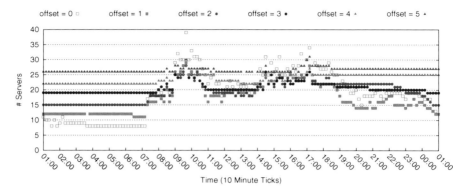

Fig. 6.10 Number of active servers on a typical day with a varying replication factor

Table 6.3 Daily server cost with varying offset

Offset	0	1	2	3	4	5
Cost ($)	201.45	186.90	201.68	237.30	257.85	289.58
Servers (max)	39	28	27	27	28	31

number of replicas. In this experiment, we evaluate the simplest one: varying the number of replicas per tenant by adding a fixed offset to the minimum number of replicas. In this experiment, we varied the offset between 1 and 5. Figure 6.10 shows that a higher replication factor decreases the variance in the active number of servers over the day. The reason is that tenant size (or memory consumption) becomes more and more important as the number of replicas per tenant increases, up to a point where tenant load is no longer the dominating factor. Surprisingly, we find that the maximum number of servers required during peak load decreases drastically as the offset increases. Conversely, during times of low load, a high offset increases the number of active servers. For our Wednesday, an offset of four is best during peak load and an offset of zero is best when load is at its lowest level. There are stages in between, where offsets of two and three deliver the best results in regard to the number of active servers. During peak load, even with an offset of five, robustfit-inc. finds much better solutions than in its standard configuration (i.e. an offset of zero).

Table 6.3 shows that cost for server rent does not increase monotonically with a higher replication offset. In fact, increasing the offset from zero to one decreases cost from $201 to $187. As a point of comparison, running portfolio-inc.—supposedly the best incremental heuristic—with an offset of zero accounts for a daily cost of $192 (see Table 6.1).

Figure 6.11a shows the average number of temporarily overloaded servers per 10 min interval for a varying offset (cf. Experiment (iii)). The standard deviation from the average value is also shown (using error bars). The average number of temporarily overloaded servers decreases proportionally with increasing offset. The

6.2 Evaluation of Algorithms for RTP

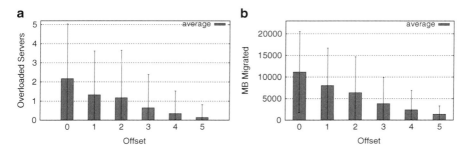

Fig. 6.11 Temporarily overloaded servers and MB migrated with varying offset. (**a**) Average number of temporary overloads. (**b**) Average amount of MB migrated

same holds true for the standard deviation. A low standard deviation suggests that the majority of values is centered around the average. The higher the number of replicas per tenant, the smoother the distribution of excess load in the cluster. Consequently, a placement becomes more and more robust towards load changes between two 10 min intervals as the number of replicas per tenant is increased.

Figure 6.11b shows the average amount of tenant data that is migrated between two 10 min intervals. The standard deviation from this average is also depicted (again, using error bars). We observe that fewer data is migrated with increasing offset. Thus, in situations where few migration budget is available or reorganization intervals are short, increasing the number of replicas per tenant is helpful.

The increased stability of the placements in terms of the number of servers required, the number of temporarily overloaded servers, and the amount of data being migrated between two 10 min intervals that comes with increasing offsets can be explained as follows. At a low number of replicas per tenant, the load dimension cap_ℓ of a server is exhausted prior to its DRAM dimension cap_σ as tenants are added to a server. The load of a tenant is equally distributed across all its replicas. Thus, the load per replica decreases with an increasing offset and, in consequence, sharp changes in load are less dramatic (i.e. they have a smaller impact on the total load on the servers). In contrast, the full amount of DRAM is allocated on each server for each additional replica of a tenant. Therefore, from a bin-packing perspective, there is a shift from tenant load cap_ℓ to tenant size cap_σ being the dominant resource that is consumed on the servers as the offset increases. Since the size of a tenant changes only slowly over time in relation to its load, which dynamically varies across the day, our algorithms have to make fewer adjustments to a placement between adjacent 10 min intervals.

6.2.3 Generic Over-Provisioning Strategies

In the following, we consider generic measures for reducing the number of temporarily overloaded servers (cf. Experiment (iii)) as well as scenarios in which more than one server fails. It turns out that the best strategy to avoid overloaded servers also helps when dealing with multiple server failures.

Two strategies immediately come to mind for reducing the number of overloaded servers: (i) virtually increasing the load of each tenant, and (ii) increasing the headroom that is left unused on each server. We simply multiply the tenants' load by a scale factor to realize the first strategy and decrease the server capacity by a constant for realizing the second. Experiment (vii) inspires a third strategy: increasing the number of replicas per tenant beyond their lower bound. The intuition behind the latter strategy is that a higher replication factor could help to smooth out harsh load changes. All three strategies can be combined with any one placement algorithm.

Experiment (viii): Reducing the Number of Overloaded Servers

Since the over-provisioning strategies described above are independent of the placement algorithms, we limit ourselves to evaluating them with robustfit-inc. Figure 6.12 visualizes how the over-provisioning schemes influence the trade-off between operational cost and temporarily overloaded servers. It consists of two sub-graphs. Each point in either graph represents an aggregate value computed across all 10 min intervals of the day. Both graphs show the total cost of all placements across our exemplary Wednesday on the y-axis.

For both graphs, we vary the strength of the over-provisioning strategies along the x-axis from left to right, i.e. by increasing either (i) the headroom left unused on the servers; (ii) the load of the tenants; or, (iii) the number of replicas per tenant. Increasing any of these three parameters results in more active servers and the resulting placement becomes more expensive in turn. Instead of showing the parameters of the individual over-provisioning strategies (e.g. the offset of the number of replicas), we show the total cost for the whole day on the y-axes. Hence, cost increases from left to right in both graphs. As a point of reference, both graphs also show the static placement from Experiment (i), which was obtained with robustfit-static-interleaved, in the top right corner (blue arrows). This reference placement produces no temporarily overloaded servers because it is strongly over-provisioned.

The difference between both sub-graphs is in their x-axes. Figure 6.12a shows the sum of all temporarily overloaded servers that are observed in any of the 10 min intervals of our exemplary Wednesday. A day consists of 144 10 min intervals, hence the large total number of overloaded servers. Figure 6.12b shows the sum of all excess load on all servers across all 144 10 min intervals. Here, excess load refers to the amount of aggregate tenant load $\mathcal{L}$ (cf. Sect. 3.2.2) in excess of the servers' load capacity (i.e. $cap_\ell = 1.0$). The latter metric is particularly sensitive. Pushing its value to a minimum would thus reduce the severity of overload situations

6.2 Evaluation of Algorithms for RTP

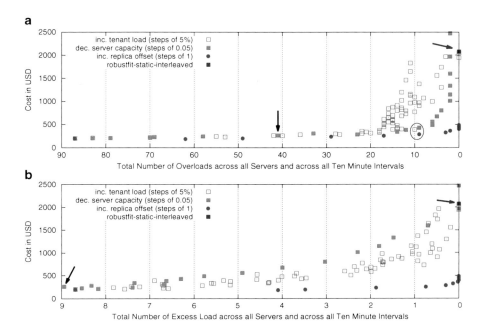

Fig. 6.12 Over-provisioning strategies for avoiding overloaded servers. (**a**) Cost of placements as a function of the number of overloaded servers. (**b**) Cost of placements as a function of total excess load across all servers

to a negligible level. For example, a point to the right in Fig. 6.12b with an x-value below 1 indicates that, across all servers in all placements for all 144 10 min intervals of the day, the total excess load is less than the load capacity of a single server, which can be considered negligible.

For both graphs, when moving from left to right along the x-axis, the resulting placements obviously become more and more expensive. When counting how many overloaded servers are observed across all 10 min intervals (Fig. 6.12a), the strategy to decrease server capacities converges towards a value of zero overloaded servers faster (and thus more inexpensively) than the strategy that virtually increases tenant load. When aggregating the amounts by which the servers are overloaded across all servers and 10 min intervals (Fig. 6.12b)—which is the more important metric for performance as perceived by the users—the opposite is the case: the strategy that virtually increases tenant load converges towards zero excess load faster.

In Experiment (iii) we have established that excess load is typically not distributed evenly among the servers in a placement; instead, most excess load accumulates on one or few servers. This can also be seen in Fig. 6.12: when decreasing server capacities, the total number of overloaded servers decreases quickly (Fig. 6.12a), while the total excess load remains relatively high as the cost of the resulting placements increases (Fig. 6.12b). In fact, the data point identified in both graphs using black arrows corresponds to the same run of the over-provisioning strategy that decreases server capacities: in this case, reducing the capacity of the servers to $cap_\ell = 0.7$ results in decreasing the number of overloaded servers to 41, while—in comparison to the original capacity limit—the total excess load increases slightly from 8.68 to 8.94.

Finally, we observe that in both graphs, and thus in terms of both metrics, the strategy to increase the number of replicas by a fixed offset is clearly superior to the other two over-provisioning strategies.

Experiment (ix): Multiple Server Failures

The penalty constraint of RTP guarantees that no server is overloaded when any one other server in the cluster fails (cf. Eq. (4.6)). Since the over-provisioning strategies introduced in the previous experiment result in placements where servers have more "headroom," we are interested in studying to what extent over-provisioned placements can absorb multiple simultaneous server failures. We study two metrics, (i) the amount by which other servers are overloaded as a consequence of one or multiple simultaneous failures, and (ii) how many tenants are rendered completely unavailable when multiple servers fail at the same time. The latter might occur when two servers fail such that both replicas of a particular tenant are rendered unavailable (in the case where this tenant has two replicas). We assume that failures occur in the middle or at the end of a 10 min interval, i.e. after all migrations have been performed. We thus collect the first metric, the excess load, *after* load changes have been observed, the placement algorithm has run, and all migrations have been performed. This is in contrast to Experiment (iii) and Experiment (viii), where excess load has been measured before the placement algorithm runs (because the focus was on the *robustness* of the placement from the previous 10 min interval). Also, in this experiment, we measure overloads only considering actual load on the servers without penalty; after all, the penalty has been designed to provide for spare capacity in failure situations, which is what we are investigating here.

We inject failures into the cluster twice during the day (marked in Fig. 6.13 using arrows). At each of these points in time, a given number of servers (between 1 and 4) fail at once. The servers within a 10 min interval that actually fail are chosen at random. For each series, i.e. for each number of simultaneous failures, we repeat the experiment 30 times using different random seeds. We compare the standard case, where no measures for over-provisioning have been applied, to the three strategies presented in the previous experiment.

For being able to compare the three over-provisioning strategies concerning their capability to deal with multiple server failures, we parameterized the strategies such that they result in similar operating costs. The configurations we picked are marked using a circle in Fig. 6.12a. For each strategy, we chose the cheapest parameterization such that less than ten overloaded servers are observed during the day.

Figure 6.13 shows the total excess load across all servers and the number of unavailable tenants for each 10 min interval of the day. We omit the night and morning hours. The points in time where failures are injected are at 11:30 a.m.,

6.2 Evaluation of Algorithms for RTP

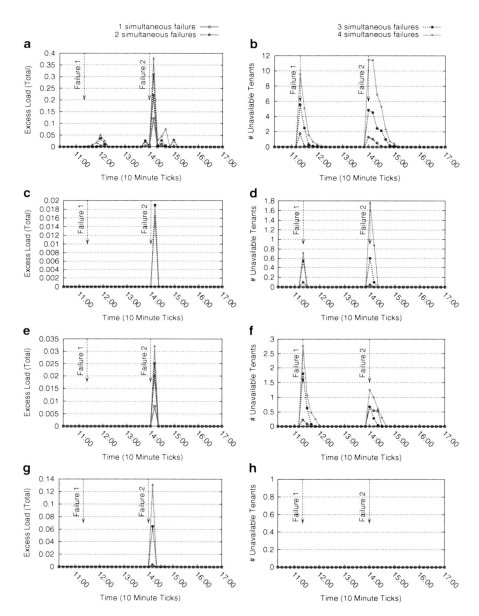

Fig. 6.13 Simultaneous server failure experiments and different over-provisioning schemes. *Left column*: total excess load. *Right column*: number of completely unavailable tenants. (**a**) robustfit-inc. (standard configuration). (**b**) robustfit-inc. (standard configuration). (**c**) robustfit-inc. with server load limit of 0.45. (**d**) robustfit-inc. with server load limit of 0.45. (**e**) robustfit-inc. with tenant load scaled by 1.85. (**f**) robustfit-inc. with tenant load scaled by 1.85. (**g**) robustfit-inc. with five additional replicas. (**h**) robustfit-inc. with five additional replicas

where load is stable after the morning spike, and 02:00 p.m., in the middle of the load increase after lunch. In the following, we discuss the results in detail.

Excess Load

At both points in time when the failures occur, no server becomes overloaded immediately (even with four simultaneous failures). The overloading in consequence of the failure occurs in the 10 min intervals after the failure. During those 10 min intervals, servers are overloaded because most of the migration budget is used up for restoring the failed replicas and thus less migration budget is available for reacting to load changes, since the priority of creating missing replicas is higher than repairing overloaded servers. If our reorganization interval were longer, say 30 min or more, and thus more migration budget was available, then the multiple failures would probably not even be visible (from looking at excess load), even without using one of the over-provisioning strategies. In our case, however, the failures cause excess load in the 10 min intervals following the failure.

The left column of Fig. 6.13 shows the amount of excess load across all servers in the cluster for the three over-provisioning strategies in conjunction with robustfit-inc. as well as robustfit-inc. in the normal configuration without over-provisioning as a point of comparison (Fig. 6.13a). The first important observation is that, for one simultaneous failure, no excess load occurs in all four cases, and especially not in the standard configuration. This is not surprising, since the penalty constraint of RTP explicitly ensures this behavior. However, the consequences of failures mainly extend to the 10 min intervals following the failure (as explained above). For the case where a single server fails, there is a moderate spike at 02:10 p.m. in the standard configuration.

In all four configurations (i.e. the standard configuration and the three different over-provisioning strategies), the failures at 02:00 p.m. cause much more excess load than the failures at 11:30 a.m. This reasonable as overall load increases gradually between 01:00 and 03:00 p.m., whereas overall load is stable during the first set of failures. In fact, for the three configurations with over-provisioning, excess load can only be observed for the second set of failures. In all three cases, this excess load occurs in a single 10 min interval.

The average amount of excess load observed when the replication-based over-provisioning strategy (Fig. 6.13g) is the second highest after the standard configuration. We observe no excess load for the case of a single server failure (in contrast to the standard case). For two simultaneous failures, there is very little excess load. For three and four simultaneous failures, we observe between four and five times less excess load than in the standard configuration.

The strategy that virtually increases tenant load (Fig. 6.13e) is second best in terms of average excess load. However, a single server failure already has a measurable impact, in contrast to the replication-based strategy. The amount of excess load observed for the strategy that increases tenant load is approximately a factor of 10 lower than for the standard case.

6.2 Evaluation of Algorithms for RTP

Virtually decreasing server capacities (Fig. 6.13c) is the most effective strategy for limiting excess load in the presence of multiple server failures. A measurable impact is only visible for three and four simultaneous failures. The severity in these cases is approximately 20 times lower than for robustfit-inc. without over-provisioning.

Note that the amounts of excess load reported in the figure are totals across all servers in a placement for the given 10 min interval. When using either of the three over-provisioning strategies, the amount by which servers are overloaded as a result of multiple simultaneous failures can be considered very moderate, given the probability that such events occur in practice.

Unavailable Tenants

The right column of Fig. 6.13 shows the number of tenants that are temporarily rendered unavailable when injecting multiple simultaneous failures into the cluster. As stated above, we vary the random seed for choosing the servers that fail and thus report average values. We make two general observations: in comparison to our analysis of excess load, (i) both sets of failures are visible also for the over-provisioned configurations; and (ii) the failures are visible immediately when they occur and, in most cases, the problem extends over multiple consecutive 10 min intervals. Also, for a single failure, no tenant is completely unavailable in all four configurations, since the minimum number of replicas in RTP is $r(t) = 2$.

In the standard configuration without over-provisioning, both the average number of unavailable tenants and the duration of the unavailabilities increases with the number of simultaneous failures. The high point is marked by an average of 11.5 unavailable tenants at 02:00 p.m. with four simultaneous failures. We observe that it takes until 03:00 p.m. to restore availability of all tenants.

The strategy that virtually increases the load of each tenant (Fig. 6.13f) produces the second highest number of unavailable tenants after the standard configuration. This is consistent with its ranking in terms of excess load. Here, this strategy requires 30 min to restore the availability of all tenants.

The strategy that virtually decreases server capacities (Fig. 6.13d) is similar to the strategy that increases tenant load concerning unavailable tenants. However, it produces shorter periods of unavailability and the amount of unavailable tenants is smaller for the set of failures occurring at 11:20 a.m.

With the replication-based strategy, not a single tenant becomes unavailable for any tested number of simultaneous failures. This comes as no surprise, since there are always more replicas than simultaneously failing servers.

Table 6.4 summarizes the performance of the over-provisioning strategies with respect to the number of unavailable tenants by listing the maximum values shown in the right column of Fig. 6.13. For all cases, the average number of unavailable tenants can be considered low, given that the total number of tenants in the experiment is 522. From an availability point-of-view, the replication-based

Table 6.4 Maximum number of unavailable tenants (averaged over 30 runs)

Simultaneous failures	0	1	2	3	4
No over-provisioning	0.00	0.00	1.75	5.59	11.50
Over-prov. strategy:					
$\ell(t)$ scaled by factor 1.85	0.00	0.00	0.68	1.80	2.78
cap_ℓ reduced to 0.45	0.00	0.00	0.10	0.60	1.76
$r(t)$ increased by 5	0.00	0.00	0.00	0.00	0.00

strategy is the clear winner. It is also the cheapest among the three over-provisioned configurations.

We conclude that although RTP guarantees that performance SLOs are met only in the presence of a single failure, in practice, multiple simultaneous server failures are often not problematic, especially when using an over-provisioning strategy.

A synthesis of the different experiments presented in this chapter will be provided in Sect. 8.2. In particular, it will be discussed what combination of algorithm and over-provisioning strategy is most effective given different priorities that a SaaS may have.

Chapter 7
Related Work

In this chapter, we survey related work in the areas of workload management and data placement. Figure 7.1 provides a high-level overview of how this dissertation is positioned among related work. The scope of the figure is limited to related work in the database community. Related approaches from other communities are presented in textual form. We discuss related work in the areas of workload management (Sect. 7.1) and data placement (Sect. 7.2) separately, following the structure of this dissertation. Some related work extends to both workload management and data placement, although there is typically a clear focus on either area. Some pieces of related work are thus discussed in both contexts.

7.1 Workload Management

The term "workload management" encompasses all processes concerned with ensuring that the database meets a user's workload objective, which can be defined in terms of throughput, response time or more complex policies. It comprises workload characterization, scheduling and admission control of queries, as well as monitoring the execution of a workload in a given system. The workload model presented in Chap. 3 of this dissertation falls into the category of workload characterization. Our survey of related work in the area of workload management is thus mainly focused on workload characterization, although we touch neighboring topics when required. Also, we focus mainly on workload characterization in database systems, and discuss workload management on application level or on infrastructure level (e.g. when provisioning virtual machines) in lesser detail.

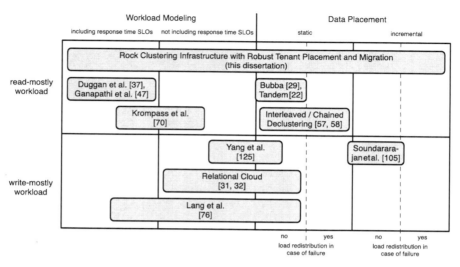

Fig. 7.1 Positioning of this dissertation among related work

7.1.1 Workload Characterization in Database Systems

In the context of DBaaS, the problem of estimating the combined resource consumption of multiple tenants sharing the same physical machine has received considerable attention over the past 2 years.

The Relational Cloud project by Curino et al. [31, 32] presents a technique for modeling the aggregate resource consumption of multiple tenants sharing the same database server. Similar to our method, they adopt an experimental rather than an analytical modeling approach. Their model is specific to the MySQL database, for which Relational Cloud develops techniques for (i) estimating the working set size inside the buffer pool and (ii) estimating the combined disk I/O performance in the presence of multi tenancy. The first problem is motivated by the observation that, after running for some amount of time, MySQL fills all memory that it can use for its buffer pool with data pages. This amount of data typically by far exceeds the DRAM required for caching all data of a tenant. Curino et al. propose a technique called "buffer pool gauging," the minimal setting for the buffer pool size can be found such that the working sets of all tenants fit into main memory. The second problem, estimating combined disk I/I, is motivated by the observation that multiple tenants with high write-rates cause a large amount of contention when trying to access the disk for the purpose of writing log files. Predicting the aggregate disk I/O performance in this case is an intricate problem, since it depends on many parameters such as log file sizes, the ratio of reads versus writes in the workload, the flushing policy for dirty pages, the number of parallel disks and their specifications, as well as the caching policies in both the operating system and the disk controller. Both problems do not occur in our setting. In in-memory column databases such as SAP HANA [42, 103], there is no buffer pool, and "paging" occurs on the level of whole columns. The working set size is simply the sum of the size of all columns (and differential buffers, cf. Sect. 3.3). Also, for in-memory databases, the problem of modeling shared access to the disk is not as central as for row-oriented databases. Disks are merely used for writing redo logs and checkpoints [123]. Also, in this dissertation the focus is on analytical and mixed workload applications with batch writes, whereas Relational Cloud focuses on write-intensive workloads such as TPC-C [112], for which the problem of shared disk access is more prevalent. Relational Cloud provides mechanisms for estimating whether a set of tenants fits on the same server in terms of aggregate resource utilization. Our workload model, presented in Chap. 3, goes further since it can be used to predict response times as a function of the degree of multi tenancy on a server.

Lang et al. [76] present a method for estimating how many tenants can be consolidated on a server before performance SLOs are violated. In contrast to our model, where response time guarantees are expressed on the granularity of servers, Lang et al. assume that tenants have different performance requirements and group them into SLO classes. Lang et al. restrict the generality of their approach by requiring that all tenants within a class have the same size. This restriction seems to be a strong one, given that the authors expect that there are merely two or three classes in total. Their model outputs whether the server is able to meet the response time goals of the different classes of tenants assigned to it. We believe that performance as perceived by the end user should be such that the experience is interactive and the user remains engaged, which can be achieved by ensuring response times below 1 s across all tenants [88]. In comparison to Relational Cloud, which models resource consumption in the presence of multi tenancy, the approach by Lang et al. takes performance SLOs into account by incorporating the response time requirement of the tenants as an attribute of the tenant classes. To this end, the model by Lang et al. has an important limitation: their guarantee that a certain average response time can be achieved for a given tenant class is based on the assumption that response times follow a Gaussian distribution. We believe that this assumption oversimplifies the problem, since the assumption does not hold for complex query workloads. In the Star Schema Benchmark [91], for example, we observed that response times follow a Gamma or Weibull distribution, depending on the particular query and tenant size [100]. Lang et al. plan to investigate how this assumption can be relaxed as part of future work. In any case, the core of their model is a binary function that returns true if the response time goals of all tenant classes on a server can be met and false, otherwise. Our model, in contrast, predicts the 99-th percentile response time for a whole server, given a set of tenants with known database sizes and request rates. Asserting a particular 99-th percentile response time is a stronger guarantee than asserting a certain average response time, since the 99-th percentile provides an upper bound for the worst case response time. Another advantage of our model is that, in addition to merely indicating whether or not a given set of tenants fits on a server, our prediction provides an intuition of how much "headroom" is left on a server before response times reach a critical level. In our graphical overview of related work in Fig. 7.1, the approach by Lang et al.

thus extends only partially into the quadrant "workload modeling including end-user performance SLOs." The model by Lang et al. has inherent support for multiple server types. This is a straight-forward extension of the model presented in Chap. 3 of this dissertation. Similar to Relational Cloud, the approach by Lang et al. relies on empirical modeling. Consequently, their model depends on the database system and workload used for conducting experiments. The study by Lang et al. is based on MS SQL Server and the TPC-C [112] benchmark. Our focus is on read-mostly workloads with batch writes. As we have already discussed in Sect. 3.6, models for either class of workload cannot easily be adapted for the other class of workload.

Duggan et al. present a mechanism for predicting query latencies in the presence of concurrent OLAP queries in [37]. Their study is based on PostgresSQL and the TPC-H [114] benchmark. Similar to our approach, their empirical model is based on a logical I/O metric which is used to characterize the dominant resource bottleneck. In their case, this I/O metric aggregates lower level metrics such as buffer pool hits and hit rates in the operating system's file system cache, which are dependent on the shared disk access behavior of concurrent queries. In our case, the corresponding metric aggregates memory I/O bandwidth consumption and CPU usage. Duggan et al. focus on predicting the response time of individual queries as they enter the queue for processing, based on the query mix that is currently being executed. The predictor is updated as queries enter and leave the system. The workload model presented in this dissertation estimates response times across all queries of a particular tenant, which is sufficient for our purposes. Additionally, we have shown that our metric for "scan capacity" consumption is additive across multiple tenants. As a result, our model can be used for predicting the increase in response times resulting from adding a new tenant to a server *before* the tenant is actually added. Duggan et al. do not discuss the additivity of their logical I/O metric across multiple tenants. Also, in contrast to our model, their work does not address the case where servers are packed with heterogeneous tenants.

All three related approaches discussed above focus solely on resource utilization that originates from processing queries. Our model also quantifies the resource consumption of administrative tasks such as tenant migrations and studies the impact that such tasks have on the response times of ongoing queries. Also, it is worth noting that the workload model presented in this dissertation was published prior to these three pieces of related work [99].

Yang et al. [125] propose to consolidate many small MySQL database instances onto a single server. They characterize the resources of a server using a vector representing CPU cycles, main memory size, and disk I/O. Each tenant consumes a fraction of each component of the vector. The size of these fractions is stored in a tenant vector. For determining the tenant vector, the tenant database is run in isolation for an unspecified amount of time and the resource consumption is monitored. A server is considered full if the sum of all tenant vectors exceeds the server vector in any dimension. This model is very coarse-grained. The authors do not provide details on the resource monitoring, which is crucial for characterizing a tenant's resource utilization. Furthermore, Yang et al. assume that all components in the vector are additive. At least for disk I/O, however, resource consumption is not similar, as was demonstrated in other work [32, 76]. Consequently, in Fig. 7.1, the approach by Yang et al. extends only partially into the quadrant "workload modeling not including end-user performance SLOs."

In the following, we discuss related workload management techniques for database systems without an explicit notion of multi tenancy.

Similar to our model, Ganapathi et al. [47] and Tozer et al. [111] use linear regression to predict the response times of individual database queries. Ganapathi et al. collect a set of static properties of a query into a feature vector. These properties can be derived from the SQL statement and information about the workload, such as the number of nested subqueries, the number and types of selection predicates, the number and types of join predicates, as well as the number of sort and aggregation columns. Their experimental model constructs a linear predictor that maps this feature vector to the response time of the query. Ganapathi et al. found that the prediction accuracy was not satisfactory when using linear regression. They thus replaced the linear regression with a machine-learning technique called "Kernel Canonical Correlation Analysis (KCCA)" [12]. In turn, they are able to predict the response times of individual TPC-DS [113] queries within 20 % of the actual response times in more than 85 % of all cases. This high accuracy comes at the expense of the high complexity of the model, which makes it hard to update when workload or hardware change. The model of Ganapathi et al. is especially useful for predicting the response times of long-running queries. In contrast, we are interested in predicting the 99-th percentile latency across all queries on a server in the presence of many concurrent queries on multiple heterogeneous data sets. Tozer et al. estimate the response times of TPC-W [115] as they enter the queue for processing based on the "query mix" currently being executed. Their goal is to use their prediction model as a component in an admission control system, which either admits or rejects incoming queries based on their predicted execution time. We are not interested in rejecting queries, but instead use our model to predict whether more tenants can be added to a server such that all queries on a server can still be executed within a fixed response time bound. The approach by Tozer et al. cannot be used for our purpose, since they assume that all queries are executed on the same TPC-W data set. Our model supports heterogeneous tenants, i.e. multiple data sets with different sizes and different numbers of requests per second executed on these data sets.

Krompass et al. [70] present an admission control and query execution framework based on fuzzy logic. Their focus is on data warehousing queries, in particular batch jobs and "interactive jobs," which are sequences of potentially hundreds of queries with a soft deadline. Their work is concerned with admission control, scheduling, and execution control of these jobs in a way that avoids overloading the database system. They consider performance SLOs based on soft deadlines in the form of clock times; there is a reward for finishing early, and a penalty for finishing after the deadline. The focus of our workload management technique is on capacity planning and automated provisioning of resources. We try to maximize throughput (i.e. by adding more tenants to a machine) until the point where our performance SLO is violated. Our performance SLO is based on interactive end-user response times rather than soft deadlines. In our classification in Fig. 7.1, the approach by Krompass et al. is on the boundary of techniques that provide performance guarantees for interactive end-user response times and those that do not.

7.1.2 Workload Characterization in Other Areas

The goal of the workload modeling technique presented in this thesis is to increase the utilization of individual servers by increasing the degree of multi tenancy on the servers. Our study focuses on multi tenancy on the database level. We use the term "workload" to denote a database schema and SQL queries that the tenants run on a shared server. Workload management is, however, not only studied in the context of databases. Another prominent domain of workload management is the consolidation of multiple applications onto a server using virtual machines. The general motivation for the studies in this area is that enterprise data centers are often under-utilized and that utilization can be increased using virtualization.

Several methods for profiling the resource requirements of virtual machines have been proposed. Urgaonkar et al. [116] collect statistics about kernel events during the execution of an application in an isolated VM. These events include system call invocations such as memory, file system and network operations. As a result, their method reports what fraction of a server's resources an individual VM consumes. Our model goes beyond indicating what fraction of a server's resources is consumed by a tenant: we predict the 99-th percentile latency of all queries handled by the server.

Zheng et al. [126] propose an automated tool that clones a virtual machine in a sandbox environment and duplicates the workload on the original VM to the sandbox VM. The resources assigned to the sandbox VM are then gradually decreased while the implications on performance in the sandbox are monitored. This allows to probe how much resources an application actually needs, which is a first step towards improving data center utilization. Zheng et al. do not study the effects that combining multiple applications as virtual machines on a server have on performance.

One aspect of the AutoGlobe project [52, 101] is concerned with the question of how multiple virtual machines can be consolidated onto a physical server without violating the performance SLOs of the individual applications running inside the virtual machines. Their approach is based on monitoring the resource consumption of the applications in terms of the CPU shares and main memory sizes required by the applications. CPU shares and memory requirements are aggregated for all virtual machines that are to be assigned to a server. The total resource requirements must not exceed the capacity of a server. Their focus is on adjusting the provisioned resources as the applications are executed, with the goal of minimizing overall SLO violation. Their approach to modeling resource consumption is time-based and predictive in nature. A synthetic trace modeling the future CPU and memory requirements of an application is generated based on historical information using a Fourier transform-based scheme. This trace is used for the purpose of consolidating multiple virtual machines onto a single server. In this dissertation, workload characterization is not done in a predictive way. Instead, our model aims to predict the response times of a database server given the *current* resource demands of the tenants, which are expressed in terms of request rates and sizes.

Gong et al. [53] propose a similar approach to AutoGlobe, but have a stronger focus on techniques for generating synthetic load traces modeling future resource demands. They develop a special-purpose algorithm for predicting the load curves, which they show to be better than auto-regression and histogram-based techniques.

7.2 Data Placement

The question of how to distribute data in a cluster of databases has a fundamental impact on the overall performance of the cluster. The data placement problem has been systematically studied in the context of distributed and parallel databases. Since most parallel database systems are designed for a shared-nothing architecture, load balancing in these systems is heavily dependent on data placement. This is in contrast to shared-memory or shared-disk architectures, where each processing node can access any piece of data [109, p. 505]. More recently, distributed storage systems for Web applications (i.e. so-called NoSQL databases) have been proposed, which also consider the data placement problem to some extent. Our goals and system model differ from the areas of distributed, parallel, and NoSQL databases. Table 7.1 summarizes these differences.

The most significant parallel database projects, i.e. Bubba [29], Gamma [35], Teradata [44, 57], and Tandem NonStop SQL [22] assume clusters with a fixed number of machines. NoSQL systems such as Dynamo [34], SimpleDB [4] (Amazon), Cassandra [74] (Facebook), HBase [49] (Apache) or PNUTS [27] (Yahoo) provide mechanisms for dynamic cluster sizing. The ability of varying the number of servers in the cluster is also a requirement for the data placement algorithms developed in this dissertation. Additionally, our approach to dynamical cluster sizing aims to minimize the number of required servers at each point in time while guaranteeing strict performance SLOs, even in the presence of server failures.

Parallel databases are designed mainly for handling large relations, which are usually *declustered*, i.e. they are horizontally partitioned into fragments that are distributed across multiple nodes in the cluster. In parallel databases, the size of those relations is often assumed to be so large that *full declustering* must be applied, i.e. the number of fragments equals the number of nodes in the cluster. To speed up query processing and increase availability, the declustered fragments are often replicated. NoSQL systems typically store their data in key/value pairs and have no logical support for the concept of a relation in the database sense. Thus, all read operations in these systems are realized on top of the lookup operation of a single key/value pair, which can be implemented efficiently using consistent

Table 7.1 Positioning of this dissertation among parallel and Web databases

Parallel databases	NoSQL databases	Rock/RTP
Fixed cluster size	Dynamically sized cluster	Dynamically sized cluster with a minimum number of nodes and guaranteed performance SLOs even in spite of failures
Fully decluster large relations and replicate fragments	Replicated key/value pairs	No large relations
Minimize response times of individual queries	Provide maximum scalability	Maximize utilization for multiple queries
Static reorganization (of relation fragments)	Incremental reorganization (of key/value pairs)	Incremental reorganization (of whole relations)

hashing [34]. Key/value pairs are typically replicated across multiple nodes in the cluster, although this replication is not performed with transactional guarantees (cf. *eventual consistency* [119]). We are interested in small relations that do not require partitioning for performance, as is the case for large relations. We replicate data on the granularity of whole relations, in a system that provides full support of the relational model and strong consistency.

In parallel databases, the goal is to minimize the response times of individual queries against the declustered relations. The main design goal of NoSQL databases is to support very large datasets by scaling to a large number of nodes. In contrast, we are interested in maximizing the number of concurrent queries on a server (by assigning multiple tenants to it) such that a fixed performance SLO can be maintained and the overall number of servers is minimal.

Finally, the data placement algorithms employed in parallel databases are not designed to support incremental changes to the layout of the fragments in the cluster: a data layout is chosen such that certain availability guarantees can be made, but is reconsidered only in relatively large intervals (e.g. weeks or months). The parallel database is typically off-line while reorganization is performed. Some NoSQL databases try to react to changing load situations in the cluster by redistributing the key/value pairs in the cluster. However, in systems based on consistent hashing, nodes are typically organized in a logical ring, and data rebalancing occurs only among neighboring nodes [34]. This limits the flexibility of NoSQL systems when reacting to overload situation, since reconfigurations that distribute the excess load on an individual node across all nodes in the cluster are not possible. Our approach entails migrating on the level of whole relations (or copies thereof) in a way that is not detrimental to performance.

Given the lack of schema enforcement and standard database functionality (such as joins) in NoSQL databases, as well as the lack of standard interfaces such as ODBC or JDBC, NoSQL databases are not suitable for supporting the kind of enterprise applications that motivate our work. We thus focus mainly on data placement strategies in parallel databases in the following. Afterwards, we consider more recent work on data placement in distributed databases. Finally, we discuss data placement issues in non-database research areas, such as dynamic virtual machine consolidation.

7.2.1 Placement Strategies in Parallel Databases

As already stated above, partitioning or declustering is concept found in most parallel databases. The goal is to horizontally partition large relations across multiple nodes so as to provide for maximum inter- and intra-query parallelism. The challenge is to find the optimal number of partitions so as to balance response times with increasing communication costs (e.g. for distributed join processing) as the number of partitions increases. The declustered fragments are often replicated and placed across the nodes in the cluster in a way that improves data availability, which we shall discuss briefly in the following. Afterwards, we will provide a detailed overview of the most important declustering techniques found in parallel databases.

7.2.1.1 Techniques for Improving Data Availability

Existing techniques for improving data availability can be grouped into three categories: identical copy approaches, inverted file strategies, and error correction techniques.

Availability strategies based on the identical copy approach involve maintaining an up-to-date copy of the primary data set in the same format as the primary copy. The advantage is that, in case of failure, the secondary copy is immediately available for query processing. Also, identical copy approaches can be combined with an active/active load balancing scheme across both copies. Naturally, maintaining two identical copies requires twice the space of the primary data set.

The inverted file strategy was designed for the Bubba parallel database [29]. The primary copy—called the *direct copy*—stores the base relation of a table, whereas the second copy—termed the *IF copy*—stores inverted indexes plus the "remainder relation" containing all non-indexed attributes. In case either copy becomes unavailable, one has to re-create the direct copy from the indexes and the remainder relation or vice-versa. This is problematic since this recovery operation is computationally expensive and negatively affects the performance of ongoing queries. The advantage of the inverted file strategy is that point queries with equality predicates on indexed attributes can be performed efficiently [29]. It is important to note that an identical copy strategy could be implemented such that indexes are stored and maintained along with each copy. In doing so, point queries could also be executed efficiently although more space would be required for storing the indexes.

Error correction techniques can be implemented in the storage layer underneath the database system. Examples are Synchronized Disk Interleaving (SDI) [67] or RAID [50, 92], which interleave sequential blocks of data (e.g. disk sectors) across multiple disks and store check bytes and parity bytes for each block.[1] In case of a disk failure, the corrupted bytes can be restored using a bit-wise XOR comparison

---
[1] While Patterson et al. [92] define multiple levels of RAID, we restrict our discussion to RAID-5.

with the check bytes. In contrast to SDI, the check blocks are spread across multiple disks in the RAID approach, which improves throughput. Parity blocks cannot be directly used for processing read requests. As with the inverted file strategy, disk sectors are unavailable during recovery. However, error correction techniques require less disk space than identical copy approaches.

In this dissertation, the focus is on availability techniques that require no downtime during recovery, since we require that our response time goal is met even in spite of a single node failure. Given the increase of storage and main memory capacities over the last decade, identical copy replication becomes increasingly attractive since identical copies can be directly used for servicing read requests. Therefore, we will restrict our discussion to identical copy approaches in the following.

7.2.1.2 Placement of Declustered Fragments

In the parallel databases Teradata [57], Gamma [35], Bubba [29], and Tandem Non-Stop SQL [22], the data placement problem entails distributing a static collection of relations across a fixed-size cluster of servers so as to minimize response times. As stated above, communication cost increases together with the number of partitions a relation is split into. The reason is that some database operations, such as joins, require shipping data or intermediate results between nodes. This communication overhead can have detrimental effects on the throughput of a parallel database system [29, 109].

In shared-nothing systems, load balancing is deeply tied in with the distribution of partitions across servers. One has to ensure that the load is balanced across partitions. In Bubba, partitions are placed in decreasing order of their access frequency, or "heat." At each placement step, the algorithm tries to balance the overall heat at each server. The problem with using the number of accesses to a partition as a metric for utilization is that it does not take query complexity into account. Our metric for utilization (Sect. 3.2.2), estimates the actual load based on access frequency.

Parallel databases typically maintain two copies of the data to ensure high availability. For small relations, a common approach to data placement is to simply treat each copy as a separate relation, with the additional constraint that the two copies cannot be placed on the same server. Bubba, for example, maintains the copies in different formats (i.e. direct and IF-copy) and tracks the heat of each copy independently. The data placement method of Bubba does not ensure that, upon the failure of any node, load is redistributed evenly across the remaining nodes. The same holds true for the Tandem parallel databases. This limitation is also visualized in Fig. 7.1. In the following, we present existing work concerned with such redistribution of load in the failure case.

For large relations, Teradata uses a technique called interleaved declustering [57]. Each of the N servers in a cluster is made responsible for the primary copy of one fragment of a relation. The secondary copy of each fragment is divided into $N-1$

7.2 Data Placement

Table 7.2 Interleaved declustering with four nodes

Node:	n_0	n_1	n_2	n_3
Primary:	F_0	F_1	F_2	F_3
		f_0^0	f_0^1	f_0^2
Secondary:	f_1^2		f_1^0	f_1^1
	f_2^1	f_2^2		f_2^0

sub-fragments that are distributed across the other servers in the cluster. Thus, when a server fails and the primary copy of a fragment becomes unavailable, the work is redistributed across $N-1$ other servers. Table 7.2 shows an example with four fragments $F_0 \ldots F_3$ and $N = 4$ nodes. We assume that all four relations receive the same load. The primary copy of F_0 resides on n_0. The secondary copy of F_0 is split into three sub-fragments f_0^0, f_0^1, f_0^2 which are assigned to nodes n_1, n_2, n_3, respectively. In case node n_0 fails, the load of F_0 is balanced across $n_1 \ldots n_3$. One disadvantage of interleaved declustering is that fragments, which are already partitioned, must be partitioned further to obtain the sub-fragments forming the secondary copy of a fragment. This must be done in a way that all sub-fragments receive approximately the same load in case of failure. Another disadvantage is that, if any two servers in the cluster become simultaneously unavailable, two relations become unavailable, since two primary fragments and two secondary fragments of the relation are off-line. This problem becomes more severe with increasing cluster size, since there are more servers than can fail at any point in time. Another downside of interleaved declustering is that maintaining consistency of the secondary copy of a fragment is expensive because of the sub-fragmentation across $N-1$ nodes.

As an alternative, Gamma uses a technique called chained declustering [58] in which the servers are organized into a logical ring. Each of the servers is made responsible for the primary copy of one fragment and the secondary copy, which is not sub-fragmented, is placed on the successor node in the ring. Again, relations are fully declustered. In the non-failure case, all load is directed towards the primary copy of the fragments, i.e. the secondary copies are run in hot standby mode. If a server fails, its load is taken over by its successor in the ring, which seemingly leads to an unbalanced system. This imbalance is avoided by changing the load balancing scheme for the entire cluster. This is done by including the secondary copies of the fragments in the load balancing in a way that the load on each node increases only by a factor of $\frac{N}{N-1}$. Table 7.3a shows an example of chained declustering during normal operations with four nodes. The corresponding failure scenario is shown in Table 7.3b. The advantage of chained declustering over interleaved declustering is that it can sustain two simultaneous node failures without a fragment becoming unavailable, unless the two nodes are adjacent in the logical ring structure. The availability rationale of chained declustering is based on the observation that the probability that two adjacent nodes fail is much higher than the probability that *any* two nodes fail [58]. Chained declustering essentially shifts the problem of load redistribution in the case of failures to the load balancing layer. In fact, the

Table 7.3 Chained declustering with four nodes

Node:	n_0	n_1	n_2	n_3
(a) Non-failure case				
Primary:	F_0	F_1	F_2	F_3
Secondary:	F'_3	F'_0	F'_1	F'_2
(b) After failure of node n_0				
Primary:	—	$F_1(\frac{1}{3})$	$F_2(\frac{2}{3})$	$F_3(1)$
Secondary:	—	$F'_0(1)$	$F'_1(\frac{2}{3})$	$F'_2(\frac{1}{3})$

load balancing scheme for the failure case is fairly complex, since it depends on the failing node and requires that the load balancer can be configured to distribute queries to fragments on such a fine-grained level. If such fine-grained control over the load balancer is not available, chained declustering degrades to a mirroring approach and nodes must be over-provisioned by 100 % to sustain additional load coming from the failure of another node.

In the following, we discuss how the data placement concepts developed in this dissertation differ from interleaved and chained declustering. Declustering mechanisms are fairly static in the sense that removing or adding a new node to the cluster is likely to require all nodes to perform work towards reorganizing the placement to deal with the new situation. Watanabe and Yokota [120] have proposed an extension of chained declustering which tries to address load imbalances that arise as a consequence of load changes. Their approach, called adaptive overlapped declustering (AOD), entails splitting the secondary copy into two disjoint sub-fragments, which are placed on the predecessor and the successor node of the location of the primary partition, respectively. The left and the right sub-fragment of the secondary copy need not have the same size. By changing the key at which both fragments are split, and by using both the primary and secondary copy for request processing, AOD can react to small fluctuations. This split point can be adjusted dynamically by migrating data between the two sub-fragments of a secondary copy. This mechanism allows to defer a global reorganization of the placement but does not alleviate the fact that global reorganizations are necessary from time to time. Our algorithms make frequent, local changes to the placement to deal with changes in tenant load which are observed in short intervals, with the goal of using a minimal number of nodes at each point in time. Therefore, if at all, declustering algorithms should only be compared to static tenant placement. However, in static RTP, the placements are more flexible from a structural perspective (i.e. nodes need not be organized in a ring and there is no notion of neighboring servers) than in data placement strategies using declustering.

Given our focus on small- to mid-size businesses, the database tables of a tenants are so small that—using current server hardware—there is no benefit to static partitioning, as required by all declustering algorithms. Thus, we consider tenants as atomic units. Placement and replication are performed on the level of whole relations. In this setting, both interleaved and chained declustering cannot be applied directly, even when dropping the requirement that relations are partitioned and thus assuming that a tenant corresponds to a fragment in the declustering sense. Such

7.2 Data Placement

a mapping of tenants to fragments would require that all tenants have equal load, since declustering algorithms require that fragmentation is done in a way that load is balanced among all fragments. From our analysis of one of SAP's on-demand applications (cf. Sect. 2.4), we know that tenants are very heterogeneous in load. Also, the number of replicas per tenant is sometimes greater than two and varies on a per-tenant basis. Declustering mechanisms do not consider varying replication factors across fragments.

Our tenant placement algorithms do not require that the load balancing mechanism can be configured on the granularity of individual relations or partitions. Instead, our algorithms are designed to work with simple round-robin load balancing. When used in conjunction with round-robin load balancing, interleaved declustering redistributes the load in case of a failure across $N - 1$ replicas. Assuming that fragments in the declustering sense can be interchanged with tenants in our sense, interleaved declustering could thus be used to devise a static tenant placement strategy that offers load redistribution. However, in the failure case, it would not be possible to enforce our performance SLO since our metric for tenant load does not model the cost of distributed join processing. Given that our tenants are not partitioned, there are no distributed joins in our approach. However, given the sub-fragmentation of the secondary copy in the interleaved declustering approach, distributed joins would become necessary when using interleaved declustering for devising a tenant placement strategy.

To implement a tenant placement strategy based on chained declustering in a setting where the load balancing scheme cannot be adjusted dynamically, one could use static recovery lists (which are used in scheduling algorithms for fault-tolerant multi-processor systems) that specify how load is redistributed in case a given processor fails. Klonowska et al. [68] present a technique for generating such recovery lists in a way that load is equally distributed across nodes in the case of a failure. Their technique is based on optimal Golomb rulers. A Golomb ruler is a set of marks at integer positions along an imaginary ruler such that no two pairs of marks are the same distance apart. Computing these marks is so complex that solutions are only known for rulers with up to 26 marks [84]. Consequently, the technique by Klonowska et al. currently only scales to recovery lists for up to 26 processes (or, in our case, declustered fragments or tenants).

From an availability perspective, the approach of interleaving tenant replicas presented in Sect. 4.1.1 of this dissertation is preferable over interleaved declustering: in contrast to interleaved declustering, two simultaneous failures do not necessarily result in any tenant becoming unavailable (when assuming a replication factor of two replicas per tenant). The probability that two simultaneous failures render a data set unavailable is the same for our interleaving approach and chained declustering (see related discussion in Sect. 4.1.1, esp. Eq. (4.9)).

Many improvements of chained declustering have been proposed, such as group rotational declustering [24], graduated declustering [10], shifted declustering [127], and adaptive overlapped declustering [120]. The limited applicability of chained declustering to our setting also applies to all of these techniques. We therefore abstain from discussing them in detail.

7.2.2 Recent Approaches for Database Replica Placement

In this section, we review related work in the area of replica placement in database clusters outside of the parallel database context.

To our knowledge, Bernstein et al. [14] use interleaving to assign database replicas in Microsoft SQL Azure. However, neither details on the design of their algorithm nor a study of its effectiveness are available.

The approach by Yang et al. [125], which we have already discussed above in the context of workload modeling, also entails assigning tenants to servers in a landscape of MySQL instances. Similar to our approach, tenants are not partitioned and their tables are placed as a whole. Yang et al. propose to perform tenant placement as follows. New tenants are monitored in isolation to determine their performance vectors (cf. Sect. 7.1.1). Afterwards, they are added to the cluster of shared databases using a variant of the on-line first-fit algorithm. Replicas are not treated in a special way, thus there is no explicit mechanism that facilitates load redistribution in case of node failures. The authors state that a technique for determining a layout for all tenants at once (rather than only adding new tenants as they join the system) is more difficult and left for future work. Our static placement algorithms do just this. In Fig. 7.1, the approach by Yang et al. extends only partially into the quadrant "static data placement without load redistribution in case of failure." Yang et al. do not consider incremental reorganizations in response to changing tenant demands.

Both Relational Cloud [32] and Lang et al. [76], also discussed in Sect. 7.1, provide simple mixed integer programs for tenant placement. They only consider the static variant of the placement problem and in both cases there is no notion of either interleaving or load redistribution in case of failures. As we have shown in Experiment (v), the running times of mixed integer programs are too high for problem sizes of practical relevance. We therefore present a suite of heuristics and randomized algorithms (as well as hybrids of both) in Chap. 5.

Supporting live data migration has been recognized as an important capability to enable multi tenancy. Elmore et al. propose Zephyr [38], a system for on-line tenant migration in shared-nothing databases with the design goal of incurring as little overhead as possible during migration. Their focus is on efficiently implementing the migrating process to support OLTP workloads. They discuss various fault-tolerance properties of their migration technique. The authors state that the motivation for Zephyr is to elastically size the cluster based on the current demand. However, the placement problem itself is not addressed. Zephyr could be used as a replacement for the simpler migration approach presented in Sect. 3.4.1 of this dissertation by adapting the factors by which migrations reduce the capacity of the source and destination servers of a migration.

Soundararajan et al. [105] propose a technique for dynamically adding and removing database replicas in a setting where multiple applications share a cluster of databases. Such addition or removal of replicas is done in response to changing load requirements of the applications. In times of low load, a particular database

7.2 Data Placement

may not be replicated at all, while in times of high load, multiple replicas may be necessary to maintain performance SLOs. Soundararajan et al. acknowledge the fact that migrating on-line replicas is expensive and has a negative effect on the response times of queries that are processed against a database server while a migration is in progress. To reduce the impact of migrations, they introduce a technique called "partial overlap," where a tenant has a primary and a secondary set of replicas. The replicas in the primary set are active: load is balanced across them and writes are synchronously applied. The replicas in the secondary set are inactive, and they are synchronized with the primaries in periodic intervals using batch writes. Secondary replicas are co-located with primary replicas of other tenants. When an increase in the load of a tenant necessitates the creation of a new replica, a replica from the secondary set is activated and promoted to the primary set. This results in fast provisioning and the impact of migrations on response times can be minimized. The authors report that SLO violations can be avoided in 92 % of the time when using the partial overlap technique. The downside of this approach is that new replicas can only be activated on servers that already hold a secondary replica of the same tenant. Maintaining a high number of secondary replicas results in high costs for synchronizing these replicas within certain staleness bounds. Our approach to tenant migration is different. The workload model in Sect. 3.4.1 characterizes the resource utilization of migrations, which allows us to migrate replicas between arbitrary nodes as long as the participating nodes have enough spare capacity. In doing so, on-line migration never results in a violation of performance SLOs.

Our tenant placement algorithms are reactive in nature, which can cause temporarily overloaded servers, as discussed in detail in Sect. 6.2.1. We presented several generic over-provisioning strategies that reduce the impact of temporary overloads to a negligible level. Soundararajan et al. [105] suggest an alternative approach. Instead of over-provisioning, they present a so-called "delay-aware" strategy for provisioning replicas. The main idea is to react aggressively when load increases (i.e. the creation of new replicas is triggered relatively fast), and, in contrast, to react defensively when load decreases (i.e. the removal of replicas is delayed until it is clear that the decrease in load is not a temporary oscillation but a general trend). From the perspective of resource utilization, it is not clear whether our generic over-provisioning strategies or the delay-aware approach by Soundararajan et al. is preferable: delaying replica removal essentially results in over-provisioning, similar to the approach adopted in this dissertation. We have provided a comparison of cost for our generic over-provisioning strategy. A similar study is not available for the delay-aware method by Soundararajan et al.

A third way to minimize temporary overloaded servers is to predict the future resource requirements of the tenants and run the placement algorithm on the predicted load for the next reorganization interval rather than on the actual load. Such prediction could be done based on statistics of tenants' resource requirements from the recent past. Initial ideas for such methods have been proposed by Zhu et al. [128], although the accuracy of their predictions is not very high. Gmach et al. [52] also propose load prediction techniques which may be applicable in this context. Once prediction methods for tenant load become available, they can easily be integrated into the placement algorithms presented in this dissertation by replacing the function $\ell(t)$, discussed in Sect. 4.1, with a function for predicting tenant load.

7.2.3 Data Placement in Other Areas

Problems similar to the data placement problem studied in this dissertation are also known outside of the database community.

The so-called File Allocation Problem (FAP) is concerned with assigning individual files to the nodes of a computer network. In FAP, the availability of a cost model is assumed that characterizes the cost of storing a file f_i on node a n_j, as well the cost for updating f_i on all nodes to which it is assigned and the cost of shipping files between nodes. The rationale behind the latter cost component is that access is cheap if a node accesses a file which is located on a local disk. Otherwise, there is some communication cost associated with accessing the file on a remote location. This problem has been extensively studied in the literature, and many different cost models using different objectives for optimization have been proposed. A survey providing a qualitative comparison of these models can be found in [36].

As part of a larger research project by HP Labs on storage service providers (SSP), Anderson et al. [7, 8] study the optimization problem of assigning "workloads" (i.e. files and performance requirements for accessing these files) to storage nodes in a way that best meets the capacity and I/O performance requirements associated with these files. They observe that bin-packing algorithms cannot be applied to their case, since the I/O performance in storage systems is not additive.

Another extension of the FAP is the replica placement problem (RPP) [64, 96], which arises in content delivery networks. As an example, consider an Internet service provider who wants to replicate multimedia data (e.g. videos for streaming) in geographically distributed cache locations in the network topology. Here, the cost of remote accesses is extended by the notion of the distance to the closest node that holds a copy of the requested object. Karlsson et al. [65] survey and compare known formalizations and solutions of RPP.

Assignment strategies for virtual machines share many aspects with data placement for DBaaS. Several groups have the considered the problem of how to consolidate multiple virtual machines onto fewer physical machines. Approaches for both *static* and *dynamic* virtual machine consolidation have been proposed [15, 83, 106, 118, 122], similar to static and incremental RTP (cf. Chap. 4). As far as dynamic consolidation is concerned, most work tries to adjust the number of active servers while avoiding to violate a certain SLO. This adjustment is done by migrating virtual machines between physical servers. The algorithms for dynamic assignment are mostly based on well-known heuristics for bin-packing. Given our goal of redistributing load in the failure case, standard bin-packing algorithms cannot be used for solving RTP. We are not aware of any work on virtual

7.2 Data Placement

machine consolidation that considers replication and redistribution of load after server failures.

Ferreto et al. [43] present a linear program and heuristics to control the migration of virtual machines between servers. In addition to the pieces of related work discussed above, they study the trade-off between the number of active servers and the number of migrations that must be performed. It might be the case that a set of virtual machines could be placed using one server less than in the current assignment, but the amount of migration required to put the new assignment into place might be uneconomically high. Our formulation of RTP contains the explicit notion of a migration budget, which is an adjustable parameter of our algorithms. Anderson et al. [9] provide an excellent overview of the *data migration problem*, which is concerned with the scheduling of multiple migrations to the maximum parallel extent under time constraints. This is itself an $\mathcal{NP}$-complete problem. It is orthogonal to our algorithms, which compute what migrations shall be conducted but not how they should be scheduled.

The AutoGlobe [52] project, which has been mentioned above in the context of workload modeling, also studies dynamic virtual machine consolidation. Similar to the approach of Ferreto et al., they take migration cost into account. Besides a linear program and greedy heuristics, they present a genetic algorithm based on the multi-criterial optimization algorithm by Deb et al. [33]. Instead of combining the objectives of (i) minimizing the number of required servers and (ii) minimizing the amount of migration in a combined cost function (which naturally results in a compromise), they exploit the genome representation of a problem instance to bound the number of migrations and focus on minimizing the number of servers in the objective function. This is done by representing a placement as a genome of length $|A|$, which is the number of applications or virtual machines that shall be consolidated. Each field in the genome contains a server ID that specifies to which server the application is assigned. In each reorganization step, the genome is randomly perturbed for a configurable amount of time. In the best case, there are multiple valid solutions after completing the random search phase. By counting the number of changed fields in the genome between the starting solution and a new solution, one can immediately derive the required number of migrations. For a given number of target servers, the solution requiring the fewest number of migrations can be selected. In RTP, reorganization is performed in frequent intervals. The length of these intervals dictates how much migration can be performed. As we have shown in Sect. 3.4.2, the duration of a migration is based on tenant size. In our case, where tenants are heterogeneous, merely limiting the number of migrations would be inaccurate: migrating one large tenant can be as expensive as migrating ten smaller tenants.

Finally, the optimization community has been considering the bin-packing problem for decades [21, 48, 117], and many variations of the problem have been studied, e.g. [39, 45, 77]. However, we are not aware of any approach that takes robustness towards individual server failures, as we consider them, into account. Again, it is the property of equal load redistribution in the failure case (or penalty, cf. Constraint (4.6)) that renders existing bin-packing algorithms unusable for RTP.

7.3 Other Aspects of Multi-tenant Databases

In this dissertation, we study workload management and data placement in the context of multi-tenant databases. However, there are other sub-problems of multi tenancy, which must be addressed in order to provide a complete DBaaS solution.

Popa et al. [95] observe that many potential users of a DBaaS service have privacy concerns towards uploading their (potentially mission-critical) business data to a third-party provider. The authors present CryptDB, which runs SQL queries on encrypted MySQL or PostgresSQL databases. CryptDB is essentially a proxy which intercepts queries and rewrites them to execute on encrypted data. This way, the database administrator cannot gain access to sensitive data. For the queries in the TPC-C benchmark [112], the average response times of MySQL are 20% worse on encrypted data and rewritten queries than for unencrypted MySQL. The proxy layer adds a significant overhead, however, since it performs encryption and decryption. From an end-user perspective, processing queries on encrypted data is approximately seven times slower than for unencrypted data.

In our SSB-MT benchmark (cf. Sect. 3.1), we have assumed that all tenant databases have the same schema. In practice, customers often require the ability to make local modifications to the schema, for example to adapt a generic piece of enterprise software to their particular industry. To fulfill this requirement, an extensibility mechanism must be in place that allows tenants to add custom attributes. In a setting where multi tenancy must be implemented using the *shared table* approach (cf. Jacobs and Aulbach [60]), extensibility and consolidation are in conflict. A common solution is to map the logical schemas of multiple tenants into the same physical schema, which consists of generic attributes [121]. Aulbach et al. [11] compare several schema-mapping techniques for this purpose. They also introduce a novel method for such mapping, which groups attributes in chunks that are often read together.

Finally, given that there are many different aspects of multi tenancy, a common methodology for evaluating multi-tenant databases has not yet been established. Kiefer et al. [66] present a framework for benchmarking multi-tenant databases. This generic framework can be used to create specific benchmarks by implementing components such as tenant-specific load drivers (to generate load variations over time) and plugging in tenant-specific instances of existing benchmarks (i.e. TPC-H with a given scale factor).

Chapter 8
Conclusions and Perspectives

In order for SaaS providers to operate their services in a cost-effective manner, while providing response time guarantees, even in the presence of failures, two key challenges must be addressed: workload management and data placement. Solutions for these challenges form the core of this dissertation. SaaS providers must balance customer satisfaction (measured in response times as perceived by the end users) and the cost for operating the service. We have shown that the potential cost savings from using an incremental data placement strategy, which consolidates multiple tenant databases onto a physical server and controls how tenant replicas are laid across multiple servers, are vast: measured in hourly rates on Amazon EC2, a provider can reduce its spending on an average working day by a factor of ten, in comparison to static provisioning for peak load requirements. On weekends and public holidays the potential savings are even more dramatic. As a consequence of consolidation, however, spare capacity on the servers decreases and vulnerability to load spikes increases, which detrimentally affects customer satisfaction. The methods chosen for workload modeling and data placement determine the provider's ability to balance this trade-off.

8.1 Summary

In this dissertation, we have introduced new methods for workload modeling and data placement in the context of in-memory column databases running an enterprise mixed workload. In the following, we will summarize the main findings and reflect on the methodologies adopted for obtaining the results.

8.1.1 Modeling Scan Capacity in a Column Store

For characterizing how the degree of multi tenancy affects the performance of multiple tenants that share a database server, we selected an empirical approach to workload modeling rather than an analytical approach in Chap. 3. Empirical modeling entails running experiments with varying parameters and generalizing the observations into a model. The database system is treated as a black box. We found that, for scan-intensive workloads, a given 99-th percentile latency across all queries on a server is exceeded when more than a certain amount of data is being scanned in a fixed period of time. Scanning consumes multiple resources on a server, e.g. bandwidth between CPU and DRAM for fetching data and CPU cycles for decompressing values and performing aggregate computations. Our results show that whether the total amount of data being scanned originates from a few large (homogeneous) tenants or many small tenants is inconsequential. An in-memory column database server has a certain scan capacity, and once this capacity is exhausted, response times deteriorate.

8.1.2 Predicting 99-th Percentile Latencies

Based on our observation that 99-th percentile response times are related to the amount of data being scanned, we constructed a function $\ell(t)$ that computes the fraction of a server's scan capacity consumed by a single tenant t. We showed that this function is additive across multiple, heterogeneous tenants on a server. We further showed that a relationship exists between the sum of the values of $\ell(t)$ for all tenants on a server and the 99-th percentile response time across all tenants on the server. We generalized this relationship into a prediction model which allows us to estimate the 99-th percentile response time of a server given the size and request rate of all tenants that are (planned to be) co-located on the server (Sect. 3.2.3). This allows for determining in advance, i.e. before adding new tenants to a server, whether the server will comply with a given performance SLO. When the aggregate load consumption of all tenants on a server is such that the resulting 99-th percentile response time is lower than 1 s, the predicted 99-th percentile value is within 10 % of the actual 99-th percentile value for the case of our SSB-MT benchmark. The better the accuracy of such a workload model, the higher the utilization level at which the provider can run its servers, since fewer spare capacity must be left unused for compensating potential prediction errors.

8.1.3 Prediction in the Presence of Perturbing Factors

We showed that the basic model for predicting 99-th percentile latencies based on the sizes and request rates of the tenants (to be) assigned to a server can be extended such that predictions are still possible in the presence of perturbing events. Examples of such events are batch writes that are executed on a server, incoming or outgoing tenant migrations, or merging the differential buffer of a column. These events reduce the scan capacity available on a server while they are in progress. We extended our prediction mechanism to capture batch writes (Sect. 3.3) and migrations (Sect. 3.4), while preserving the accuracy of the basic model.

8.1.4 Robust Tenant Placement and Migration

After investigating multi tenancy on the level of individual servers, we have established the Robust Tenant Placement and Migration Problem (RTP) in Chap. 4, which addresses multi tenancy on the level of the cluster as a whole. We posed RTP as an optimization problem that strives to assign tenants to servers in a way that servers are not overloaded (i.e. response times do not exceed a certain 99-th percentile latency; we use our workload model for this purpose) and tenants are replicated at least once. This assignment is reconsidered at frequent intervals to take variations in tenant load over time into account. Since reassignment occurs frequently, it must be done incrementally in the sense that only a limited amount of data may be migrated, and reorganization occurs while the tenant databases are on-line. To ensure that response times in the 99-th percentile remain lower than the response time goal, we use our workload model that considers migration cost. RTP addresses multi tenancy on a global level by aiming to continuously minimize the number of servers required for operating all tenants. Pushing the concept of guaranteeing a certain 99-th percentile latency even further, RTP assigns tenant replicas to servers in a way that an unexpected server outage in the cluster does not result in any of the remaining servers violating this guarantee. We implement this property by interleaving tenant replicas across servers such that excess load caused by a failure is spread evenly across as many live servers as possible. We found that this property renders existing algorithms for bin-packing unusable for RTP. We also proved the $\mathcal{NP}$-completeness of RTP in Sect. 4.3.

8.1.5 Algorithm Design

Given that RTP is a new $\mathcal{NP}$-complete problem to which existing bin-packing algorithms cannot be applied, we had no intuition on what *class* of algorithm is most promising for approaching the problem. This is in contrast to the bin-packing problem, for which greedy heuristics are known to produce good solutions [30] and are therefore often a good choice. Besides designing greedy heuristics for RTP, we also created metaheuristics that incorporate random search as well as exact algorithms in the form of mixed integer programs in Chap. 5. The latter require an unreasonable amount of computation but help to establish bounds on the optimality of solutions. For the greedy algorithms and metaheuristics, we found that computing solutions of incremental RTP can be broken down into six phases (Sect. 5.2.1), in each of which a different algorithm could be used. This led to the design of hybrid algorithms that combine the strengths of various algorithms in specific phases.

8.1.6 Algorithm Evaluation

The purpose of our evaluation in Chap. 6 was to analyze how well our algorithms react to changes in tenant load occurring as part of diurnal variations. To this end, there is a vast array of assumptions that one must formulate about the nature of such variations for mimicking a realistic scenario. We were fortunate and obtained log data for a hosted multi-tenant application by SAP, from which we could extract the diurnal variations. We have summarized the most interesting findings of our analysis of this data in Sect. 2.4. Since only a sample of all tenants using this application was made available to us for protecting the customers' privacy, we developed a bootstrapping technique for increasing our corpus of tenants in order to test our algorithms at a larger scale. We also showed what key parameters of the bootstrapping process must be tuned (and how) for balancing the resemblance of new tenants to tenants in the corpus and the proneness of new tenant traces to unnatural load spikes (Sect. 6.1.3). This methodology can be re-used by other researchers in similar situations, e.g. for devising benchmarks that include diurnal load variations.

In Sect. 6.2.1, we evaluated how well our placement algorithms perform in terms of three different metrics: (i) the operating cost for the servers required by the resulting placements over the course of a typical working day; (ii) the amount of computation required to produce a placement; and (iii) the robustness of the resulting placements towards sudden increases in tenant load. In terms of cost, we found that incremental tenant placement in general leads to a radical improvement over static tenant placement, where reassignment of tenants is seldom. For example, robustfit-static-mirror, modeled after the standard practice for replicated data placement in many companies, results in an operating cost of $3,456 for 522 tenants on an average business day. robustfit-static-interleaved, a more sophisticated algorithm for static placement, generates a bill of $2,073 in terms of Amazon EC2 server hours. In contrast, the operating cost varies between $192 and $274 across all our incremental algorithms. The average time for computing an incremental placement varies between 1.7 and 182 s among the different algorithms. The fastest algorithm, robustfit-inc., is a greedy heuristic and achieves a daily operating cost of $201. In terms of robustness towards load spikes, a computationally expensive heuristic, called splitmerge-inc., achieves the best result.

Our most important findings are that (i) our robustfit-inc. and tabu-robustfit-inc.-long algorithms find near cost-optimal solutions while requiring comparatively little computation time, and algorithms that find more inexpensive placements are prohibitively slow for most practical cases; (ii) moderately increasing the number of replicas per tenant reduces the number of required servers; (iii) using our over-provisioning strategies does not only minimize the impact of load spikes to a negligible level, but also helps masking multiple simultaneous server failures from an SLO perspective, although our model only guarantees robustness towards a single server failure; and (iv) the over-provisioning strategy based on increasing the replication factor is the winner among the presented approaches, both from a cost and an availability perspective.

In Sect. 8.2 we will navigate the trade-off between these three metrics in more detail and provide guidance on what kind of strategy should be used for different sets of priorities that data center administrators may have.

8.2 Guidelines

In the experimental evaluation in Sect. 6.2.1, we have explored the space of cost (both in terms of number of servers and computation time), temporarily overloaded servers (as a consequence of consolidating too aggressively), and resilience towards server failures. In practice, the question of how to prioritize among these dimensions depends on multiple aspects and requirements of the application.

When cost is the most important factor, then tabu-robustfit-inc.-long is the best incremental algorithm, although portfolio-inc. produces marginally more inexpensive placements. The reason is that portfolio-inc. requires 3 min on average for computing a solution, and up to 9.5 min in the worst case. When load changes are observed in 10 min intervals, then there will be cases where only few of the migrations suggested by portfolio-inc. can be physically carried out. In contrast, tabu-robustfit-inc.-long requires 20 s of computation on average and 1 min at most, which leaves enough time to physically perform the suggested migrations. Note that tabu-robustfit-inc.-long is a hybrid algorithm, in which some of the phases in the framework for incremental algorithms (Sect. 5.2.1) are performed by the greedy heuristic robustfit-inc. and some are performed by tabu-inc.-long, a randomized algorithm. We were only able to arrive at this result by not restricting our study to a particular algorithm family (e.g. greedy heuristics) but investigated the spectrum of greedy, randomized, and exact algorithms, as well as combinations of individual algorithms from these families via our framework for incremental placement algorithms. For longer reorganization intervals (i.e. in excess of 20 min), portfolio-inc. is the best choice. Orthogonally to the chosen placement algorithm, the minimum number of replicas for each individual tenant should be increased by an offset of one, since this reduces cost by 10 % on average, as shown in Experiment (vii).

When robustness towards unexpected load spikes is the most important factor, then splitmerge-inc. is the best incremental algorithm. It produces placements with the lowest values across algorithms in the three metrics percentage of overloaded servers, average excess load per overloaded server, and excess load on the worst overloaded server. On average, it required 1.5 min of computation time in our experiments, and almost 5.5 min at most. Again, this amount of computation is too excessive when reorganization intervals are short. All other algorithms produce placements which should be considered too sensitive towards sudden increases in load; they should be combined with one of the generic over-provisioning strategies presented in Sect. 6.2.3. Among those strategies, increasing the number of replicas beyond the minimum number of replicas is most effective, since it reduces the amount excess load across all servers and across the whole day to a negligible value at a significantly lower cost than the other two strategies (i.e. a factor of two times or more, depending on how aggressively robustness is optimized for), as can be seen in Fig. 6.12b. One additional replica already doubles the resulting robustness to spikes, while the total cost decreases, as we have already mentioned above.

Regarding the resilience of placements towards unexpected server outages, we observe that failures do not have much of an impact. For example, when four servers fail simultaneously, there is one single 10 min interval where all servers in a placement together are overloaded by 37 % of the capacity of a single server when using robustfit-inc., our simplest greedy placement heuristic, in its standard configuration (and a few adjacent 10 min intervals with excess load between 3 and 6 %). This is very good result but not unanticipated; after all, RTP was modeled to inherently facilitate load redistribution in case of failures. While this redistribution was designed in a way that response time guarantees can be kept in the presence of a single server failure, we observe that load also redistributes further as more servers fail. Yet, if the circumstances demand even stronger resilience against server outages, excess load in the presence of multiple server failures can further be optimized by applying a generic over-provisioning strategy. In this case, the strategy that virtually decreases server capacities masks failures most effectively. At two and more simultaneous failures, however, individual tenants might become unavailable. While the strategy to decrease server capacities reduces the number of unavailable tenants in comparison to the standard case without over-provisioning by approximately a factor of ten, the strategy that increases the number of replicas is naturally best when maximizing availability is the top priority.

Regardless of the priorities concerning the above factors, the over-provisioning strategy that builds on increasing the number of replicas per tenant is worth considering: it decreases the amount of servers required during times of peak load, it decreases cost when applied moderately, it increases robustness towards load spikes, it results in fewer migrations between 10 min intervals and thus a higher stability of placements, and it maximizes availability in the presence of (multiple) server failures. Besides the drawback that the other over-provisioning strategies are less prone to excess load in the presence of multiple server failures, which can be considered minor given that the amount of excess load in the standard case without over-provisioning is already low by absolute measures, it carries the disadvantage of limited scalability: the number of replicas per tenant cannot be arbitrarily increased. For example, the most effective range for the replication-based over-provisioning strategy in Fig. 6.12b is between three and five additional replicas. Given that tenants already have two replicas to begin with, the resulting number of replicas is high, even for OLAP standards. Operating at such high replication factors might require resorting to an eventual consistency model [34] or similar approaches [69], which is not appropriate for a large class of applications.

Based on the above considerations, SaaS providers should chose a combination from our placement algorithms and over-provisioning strategies that best matches their preferences and the requirements of their application. As a general maxim, there will be temporarily overloaded servers in consequence of unexpected load spikes when entirely prioritizing cost. Spare capacity should thus always be provided; our over-provisioning strategies allow to do this in a cost-conscious manner that preserves the dramatic cost savings of incremental placement to a large extent.

8.3 Future Work

In Chap. 3, we have shown that our workload model can be extended to capture drops in server capacity incurred by batch writes and migrations. There are a number of other conceivably useful extensions that we have not studied. These include the degradation of scan capacity during (i) *merges* between the differential buffer and the main store of a column [93, 94] and (ii) online schema modifications [11]. One could also study whether the model can be extended to capture other cluster management tasks such as backups or periodical snapshots.

Despite the fact that we have presented multiple algorithms for RTP and have conducted a detailed experimental evaluation, multiple avenues remain for future work.

In RTP, the number of replicas of each tenant is an input parameter to the optimization problem. We have shown how to compute a lower bound for the number of replicas and use the result of this computation in our placement algorithms. However, Experiment (vii) suggest that determining the number of replicas per tenant as part of solving the optimization problem could provide a reduction in operating cost of approximately another 30%. We have tried to incorporate this aspect into our mixed integer program formulations of RTP; these efforts were not successful, since the number of variables in the MIP became unmanageably high. The results of Experiment (vii) have shown that the optimal number of additional tenant replicas beyond the lower bound is dependent on the total load in the cluster. Over a typical day, there are multiple tipping points, at which the best number of additional replicas changes. In general, a low number of replicas is best at night and during times of low load, while a high number of replicas results in the fewest number of servers during peak load. One could try to model the relation between total load and the number of extra replicas per tenant (perhaps using similar techniques as in Chap. 3) as a first approximation for dynamically determining the number of replicas.

Similarly, for reducing temporarily overloaded servers, more hybrid variants of the presented algorithms could be studied. An extreme approach would be to switch the placement algorithm between 10 min intervals. For example, given that splitmerge-inc. produces placements with so little excess load during load spikes, it would be conceivable to run splitmerge-inc. when the change in load between two 10 min intervals is beyond a certain threshold, and to run robustfit-inc. otherwise. In this dissertation, we limited ourselves to studying over-provisioning strategies which are orthogonal to the placement algorithms.

Finally, our approach to incremental placement is reactive in nature. Given the strongly recurring diurnal patterns observed in our analysis of the on-demand application from SAP in Sect. 2.4, it would be worthwhile to study proactive placement techniques based on predicted tenant load. The challenge lies in predicting tenant load for a given time of day based on the analysis of the tenants' behavior in the past. Such prediction is complementary to RTP in its current form, in the sense that the load function in RTP could be replaced by a predictor.

Bibliography

1. D.J. Abadi, Query execution in column-oriented database systems. Ph.D. thesis, Massachusetts Institute of Technology, Feb 2008
2. D.J. Abadi, S. Madden, N. Hachem, Column-stores vs. row-stores: how different are they really? in *Proceedings of the ACM SIGMOD International Conference on Management of Data, SIGMOD 2008*, Vancouver, 10–12 June 2008 (ACM, 2008), pp. 967–980
3. T. Achterberg, SCIP: solving constraint integer programs. Math. Program. Comput. **1**(1), 1–41 (2009)
4. Amazon Web Services, Inc., Amazon Simpledb (Beta) (2012), http://www.amazon.com/b?node=342335011 (Online). Accessed 20 Dec 2012
5. Amazon Web Services LLC, Amazon EC2 Pricing (2012), http://aws.amazon.com/ec2/pricing/ (Online). Accessed 13 Dec 2012
6. Amazon Web Services LLC, Amazon elastic compute cloud (Amazon EC2) (2012), http://aws.amazon.com/ec2/ (Online). Accessed 13 Dec 2012
7. E. Anderson, M. Hobbs, K. Keeton, S. Spence, M. Uysal, A.C. Veitch, Hippodrome: running circles around storage administration, in *Proceedings of the FAST '02 Conference on File and Storage Technologies*, Monterey, 28–30 Jan 2002 (USENIX, 2002), pp. 175–188
8. E. Anderson, S. Spence, R. Swaminathan, M. Kallahalla, Q. Wang, Quickly finding near-optimal storage designs. ACM Trans. Comput. Syst. **23**(4), 337–374 (2005)
9. E. Anderson, J. Hall, J.D. Hartline, M. Hobbes, A.R. Karlin, J. Saia, R. Swaminathan, J. Wilkes, Algorithms for data migration. Algorithmica **57**(2), 349–380 (2010)
10. R.H. Arpaci-Dusseau, E. Anderson, N. Treuhaft, D.E. Culler, J.M. Hellerstein, D.A. Patterson, K.A. Yelick, Cluster I/O with river: making the fast case common, in *IOPADS*, Atlanta, 1999, pp. 10–22
11. S. Aulbach, T. Grust, D. Jacobs, A. Kemper, J. Rittinger, Multi-tenant databases for software as a service: schema-mapping techniques, in *Proceedings of the ACM SIGMOD International Conference on Management of Data, SIGMOD 2008*, Vancouver, 10–12 June 2008 (ACM, 2008), pp. 1195–1206
12. F.R. Bach, M.I. Jordan, Kernel independent component analysis. J. Mach. Learn. Res. **3**, 1–48 (2002)
13. H. Berenson, P.A. Bernstein, J. Gray, J. Melton, E.J. O'Neil, P.E. O'Neil, A critique of ANSI SQL isolation levels, in *Proceedings of the 1995 ACM SIGMOD International Conference on Management of Data*, San Jose, 22–25 May 1995 (ACM, 1995), pp. 1–10
14. P.A. Bernstein, I. Cseri, N. Dani, N. Ellis, A. Kalhan, G. Kakivaya, D.B. Lomet, R. Manne, L. Novik, T. Talius, Adapting Microsoft SQL server for cloud computing, in *Proceedings of the 27th International Conference on Data Engineering, ICDE 2011*, Hannover, 11–16 Apr 2011 (IEEE Computer Society, 2011), pp. 1255–1263

15. N. Bobroff, A. Kochut, K.A. Beaty, Dynamic placement of virtual machines for managing SLA violations, in *10th IFIP/IEEE International Symposium on Integrated Network Management, IM 2007*, Munich, 21–25 May 2007 (IEEE Computer Society, 2007), pp. 119–128
16. P. Bodík, A. Fox, M.J. Franklin, M.I. Jordan, D.A. Patterson, Characterizing, modeling, and generating workload spikes for stateful services, in *Proceedings of the 1st ACM Symposium on Cloud Computing, SoCC 2010*, Indianapolis, 10–11 June 2010 (ACM, 2010), pp. 241–252
17. A. Bog, H. Plattner, A. Zeier, A mixed transaction processing and operational reporting benchmark. Inf. Syst. Front. **13**(3), 321–335 (2011)
18. P. Boncz, Monet: a next-generation DBMS Kernel for query-intensive applications. Ph.D. thesis, Universiteit van Amsterdam, Amsterdam, May 2002
19. M. Burrows, The chubby lock service for loosely-coupled distributed systems, in *7th Symposium on Operating Systems Design and Implementation (OSDI '06)*, Seattle, 6–8 Nov, (USENIX Association, 2006), pp. 335–350
20. Cable News Network, Fortune 500 2012 (2012), http://money.cnn.com/magazines/fortune/fortune500/2012/full_list/ (Online). Accessed 13 Dec 2012
21. H. Cambazard, B. O'Sullivan, Propagating the bin packing constraint using linear programming, in *Principles and Practice of Constraint Programming – CP 2010 – 16th International Conference, CP 2010*, St. Andrews, 6–10 Sept 2010. Proceedings (Springer, 2010), pp. 129–136
22. P. Celis, Distribution, parallelism, and availability in NonStop SQL, in *Proceedings of the 1992 ACM SIGMOD International Conference on Management of Data*, San Diego, 2–5 June 1992 (ACM, 1992), p. 225
23. F. Chang, J. Dean, S. Ghemawat, W.C. Hsieh, D.A. Wallach, M. Burrows, T. Chandra, A. Fikes, R.E. Gruber, Bigtable: a distributed storage system for structured data. ACM Trans. Comput. Syst. **26**(2), 26 (June 2008), Article 4. doi:10.1145/1365815.1365816. http://doi.acm.org/10.1145/1365815.1365816
24. S. Chen, D.F. Towsley, A performance evaluation of RAID architectures. IEEE Trans. Comput. **45**(10), 1116–1130 (1996)
25. Citrix Systems, Inc., The Xen Hypervisor (2012), http://www.xen.org/ (Online). Accessed 13 Dec 2012
26. R. Cole, F. Funke, L. Giakoumakis, W. Guy, A. Kemper, S. Krompass, H.A. Kuno, R.O. Nambiar, T. Neumann, M. Poess, K.-U. Sattler, M. Seibold, E. Simon, F. Waas, The mixed workload CH-benchmark, in *Proceedings of the Fourth International Workshop on Testing Database Systems, DBTest 2011*, Athens, 13 June 2011 (ACM, 2011), p. 8
27. B.F. Cooper, R. Ramakrishnan, U. Srivastava, A. Silberstein, P. Bohannon, H.-A. Jacobsen, N. Puz, D. Weaver, R. Yerneni, PNUTS: Yahoo!'s hosted data serving platform. PVLDB **1**(2), 1277–1288 (2008)
28. G.P. Copeland, S. Khoshafian, A decomposition storage model, in *Proceedings of the 1985 ACM SIGMOD International Conference on Management of Data*, Austin, 28–31 May 1985 (ACM, 1985), pp. 268–279
29. G.P. Copeland, W. Alexander, E.E. Boughter, T.W. Keller, Data placement in Bubba, in *Proceedings of the 1988 ACM SIGMOD International Conference on Management of Data*, Chicago, 1–3 June 1988 (ACM, 1988), pp. 99–108
30. J. Csirik, D.S. Johnson, Bounded space on-line bin packing: best is better than first, in *Proceedings of the Second Annual ACM/SIGACT-SIAM Symposium on Discrete Algorithms*, San Francisco, 28–30 Jan 1991 (ACM/SIAM, 1991), pp. 309–319
31. C. Curino, E.P.C. Jones, R.A. Popa, N. Malviya, E. Wu, S. Madden, H. Balakrishnan, N. Zeldovich, Relational cloud: a database service for the cloud, in *CIDR 2011, Fifth Biennial Conference on Innovative Data Systems Research*, Asilomar, 9–12 Jan 2011. Online Proceedings, 2011, pp. 235–240. www.cidrdb.org
32. C. Curino, E.P.C. Jones, S. Madden, H. Balakrishnan, Workload-aware database monitoring and consolidation, in *Proceedings of the ACM SIGMOD International Conference on Management of Data, SIGMOD 2011*, Athens, 12–16 June 2011 (ACM, 2011), pp. 313–324

33. K. Deb, S. Agrawal, A. Pratap, T. Meyarivan, A fast elitist non-dominated sorting genetic algorithm for multi-objective optimisation: NSGA-II, in *Parallel Problem Solving from Nature – PPSN VI, 6th International Conference*, Paris, 18–20 Sept 2000. Proceedings (Springer, 2000), pp. 849–858
34. G. DeCandia, D. Hastorun, M. Jampani, G. Kakulapati, A. Lakshman, A. Pilchin, S. Sivasubramanian, P. Vosshall, W. Vogels, Dynamo: Amazon's highly available key-value store, in *Proceedings of the 21st ACM Symposium on Operating Systems Principles 2007, SOSP 2007*, Stevenson, 14–17 Oct 2007 (ACM, 2007), pp. 205–220
35. D.J. DeWitt, S. Ghandeharizadeh, D.A. Schneider, A. Bricker, H.-I. Hsiao, R. Rasmussen, The Gamma database machine project. IEEE Trans. Knowl. Data Eng. **2**(1), 44–62 (1990)
36. L.W. Dowdy, D.V. Foster, Comparative models of the file assignment problem. ACM Comput. Surv. **14**(2), 287–313 (1982)
37. J. Duggan, U. Çetintemel, O. Papaemmanouil, E. Upfal, Performance prediction for concurrent database workloads, in *Proceedings of the ACM SIGMOD International Conference on Management of Data, SIGMOD 2011*, Athens, 12–16 June 2011 (ACM, 2011), pp. 337–348
38. A.J. Elmore, S. Das, D. Agrawal, A. El Abbadi, Zephyr: live migration in shared nothing databases for elastic cloud platforms, in *Proceedings of the ACM SIGMOD International Conference on Management of Data, SIGMOD 2011*, Athens, 12–16 June 2011 (ACM, 2011), pp. 301–312
39. L. Epstein, A. Levin, On bin packing with conflicts. SIAM J. Optim. **19**(3), 1270–1298 (2008)
40. Eucalyptus Systems Inc., Open source private and hybrid clouds from Eucalyptus (2012), http://www.eucalyptus.com/ (Online). Accessed 13 Dec 2012
41. European Commission, Small and medium-sized enterprises (SMEs): what is an SME? (2012), http://ec.europa.eu/enterprise/policies/sme/facts-figures-analysis/sme-definition/index_en.htm (Online). Accessed 13 Dec 2012
42. F. Färber, S.K. Cha, J. Primsch, C. Bornhövd, S. Sigg, W. Lehner, SAP HANA database: data management for modern business applications. SIGMOD Rec. **40**(4), 45–51 (2011)
43. T.C. Ferreto, M.A.S. Netto, R.N. Calheiros, C.A.F. De Rose, Server consolidation with migration control for virtualized data centers. Future Gener. Comput. Syst. **27**(8), 1027–1034 (2011)
44. D. Flynn, M. Adamson, Capacity planning on a Teradata DataWarehouse, in *34th International Computer Measurement Group Conference*, Las Vegas, 7–12 Dec 2008. Proceedings (Computer Measurement Group, 2008), pp. 93–104
45. French Operational Research and Decision Support Society, ROADEF challenge 2012 (2012), http://challenge.roadef.org (Online). Accessed 13 Dec 2012
46. F. Funke, A. Kemper, T. Neumann, Benchmarking hybrid OLTP&OLAP database systems, in *Datenbanksysteme für Business, Technologie und Web (BTW), 14. Fachtagung des GI-Fachbereichs "Datenbanken und Informationssysteme" (DBIS)*, 2.-4.3.2011 in Kaiserslautern (GI, 2011), pp. 390–409
47. A. Ganapathi, H.A. Kuno, U. Dayal, J.L. Wiener, A. Fox, M.I. Jordan, D.A. Patterson, Predicting multiple metrics for queries: better decisions enabled by machine learning, in *Proceedings of the 25th International Conference on Data Engineering, ICDE 2009*, Shanghai, 29 Mar 2009–2 Apr 2009 (IEEE Computer Society, 2009), pp. 592–603
48. M.R. Garey, D.S. Johnson, *Computers and Intractability: A Guide to the Theory of NP-Completeness* (W. H. Freeman, San Francisco, 1979). ISBN:0-7167-1044-7
49. L. George, *Hbase – The Definitive Guide: Random Access to Your Planet-Size Data* (O'Reilly, Sebastopol, 2011). ISBN:978-1-449-39610-7
50. G.A. Gibson, L. Hellerstein, R.M. Karp, R.H. Katz, D.A. Patterson, Failure correction techniques for large disk arrays, in *ASPLOS-III Proceedings – Third International Conference on Architectural Support for Programming Languages and Operating Systems*, Boston, 3–6 Apr 1989 (ACM, 1989), pp. 123–132
51. F. Glover, Tabu search – part I. INFORMS J. Comput. **1**(3), 190–206 (1989)
52. D. Gmach, J. Rolia, L. Cherkasova, G. Belrose, T. Turicchi, A. Kemper, An integrated approach to resource pool management: policies, efficiency and quality metrics, in *The 38th Annual IEEE/IFIP International Conference on Dependable Systems and Networks*,

DSN 2008, Anchorage, 24–27 June 2008. Proceedings (IEEE Computer Society, 2008), pp. 326–335
53. Z. Gong, X. Gu, J. Wilkes, PRESS: predictive elastic resource scaling for cloud systems, in *Proceedings of the 6th International Conference on Network and Service Management, CNSM 2010*, Niagara Falls, 25–29 Oct 2010 (IEEE Computer Society, 2010), pp. 9–16
54. J. Gray, P. Sundaresan, S. Englert, K. Baclawski, P.J. Weinberger, Quickly generating billion-record synthetic databases, in *Proceedings of the 1994 ACM SIGMOD International Conference on Management of Data*, Minneapolis, 24–27 May 1994 (ACM, 1994), pp. 243–252
55. M. Grund, J. Schaffner, J. Krüger, J. Brunnert, A. Zeier, The effects of virtualization on main memory systems, in *Proceedings of the Sixth International Workshop on Data Management on New Hardware, DaMoN 2010*, Indianapolis, 7 June 2010 (ACM, 2010), pp. 41–46
56. J.R. Hamilton, Internet-scale data center power efficiency, in *CIDR 2009, Fourth Biennial Conference on Innovative Data Systems Research*, Asilomar, 4–7 Jan 2009. Online Proceedings, 2009, www.cidrdb.org
57. L. Hedegard, J. Dietz, The benefits of enabling fallback in the active data warehouse. Teradata Mag. Online **7**(1) (2007). http://apps.teradata.com/tdmo/v07n01/tech2tech/asktheexpert/benefitsoffallback.aspx (Online). Accessed 13 Dec 2012
58. H.-I. Hsiao, D.J. DeWitt, Chained declustering: a new availability strategy for multiprocessor database machines, in *Proceedings of the Sixth International Conference on Data Engineering*, Los Angeles, 5–9 Feb 1990 (IEEE Computer Society, 1990), pp. 456–465
59. ILOG CPLEX 12.2, Reference manual (2012), http://www.ilog.com/products/cplex
60. D. Jacobs, S. Aulbach, Ruminations on multi-tenant databases, in *Datenbanksysteme in Business, Technologie und Web (BTW 2007), 12. Fachtagung des GI-Fachbereichs "Datenbanken und Informationssysteme" (DBIS), Proceedings*, 7.-9. März 2007, Aachen (GI, 2007), pp. 514–521
61. B. Jaecksch, W. Lehner, F. Faerber, A plan for OLAP, in *EDBT 2010, 13th International Conference on Extending Database Technology*, Lausanne, 22–26 Mar 2010, Proceedings (ACM, 2010), pp. 681–686
62. D.S. Johnson, Approximation algorithms for combinatorial problems. J. Comput. Syst. Sci. **9**(3), 256–278 (1974)
63. V. Kaibel, M.E. Pfetsch, Packing and partitioning orbitopes. Math. Program. **114**, 1–36 (2008)
64. M. Karlsson, C.T. Karamanolis, Choosing replica placement heuristics for wide-area systems, in *24th International Conference on Distributed Computing Systems (ICDCS 2004)*, Hachioji, 24–26 Mar 2004 (IEEE Computer Society, 2004), pp. 350–359
65. M. Karlsson, C. Karamanolis, M. Mahalingam, A framework for evaluating replica placement algorithms, HP Laboratories Palo Alto (2002), http://www.hpl.hp.com/techreports/2002/hpl-2002-219.html (Online). Accessed 23 Dec 2012
66. T. Kiefer, B. Schlegel, W. Lehner, MulTe: a multi-tenancy database benchmark framework, in *TPC Technology Conference (TPCTC)*, Istanbul, 2012, pp. 11–18
67. M.Y. Kim, Synchronized disk interleaving. IEEE Trans. Comput. **35**(11), 978–988 (1986)
68. K. Klonowska, L. Lundberg, H. Lennerstad, C. Svahnberg, Extended golomb rulers as the new recovery schemes in distributed dependable computing, in *19th International Parallel and Distributed Processing Symposium (IPDPS 2005), CD-ROM/Abstracts Proceedings*, Denver, 4–8 Apr 2005 (IEEE Computer Society, 2005)
69. T. Kraska, M. Hentschel, G. Alonso, D. Kossmann, Consistency rationing in the cloud: pay only when it matters. PVLDB **2**(1), 253–264 (2009)
70. S. Krompass, U. Dayal, H.A. Kuno, A. Kemper, Dynamic workload management for very large data warehouses: juggling feathers and bowling balls, in *Proceedings of the 33rd International Conference on Very Large Data Bases*, University of Vienna, Vienna, 23–27 Sept 2007 (ACM, 2007), pp. 1105–1115
71. J. Krüger, M. Grund, C. Tinnefeld, H. Plattner, A. Zeier, F. Faerber, Optimizing write performance for read optimized databases, in *Database Systems for Advanced Applications,*

15th International Conference, DASFAA 2010, Tsukuba, 1–4 Apr 2010. Proceedings, Part II (Springer, 2010), pp. 291–305
72. J. Krüger, C. Tinnefeld, M. Grund, A. Zeier, H. Plattner, A case for online mixed workload processing, in *Proceedings of the Third International Workshop on Testing Database Systems, DBTest 2010*, Indianapolis, 7 June 2010 (ACM, 2010)
73. J. Krüger, C. Kim, M. Grund, N. Satish, D. Schwalb, J. Chhugani, H. Plattner, P. Dubey, A. Zeier, Fast updates on read-optimized databases using multi-core CPUs. PVLDB **5**(1), 61–72 (2011)
74. A. Lakshman, P. Malik, Cassandra: structured storage system on a P2P network, in *Proceedings of the 28th Annual ACM Symposium on Principles of Distributed Computing, PODC 2009*, Calgary, 10–12 Aug 2009 (ACM, 2009), p. 5
75. L. Lamport, The part-time parliament. ACM Trans. Comput. Syst. **16**(2), 133–169 (1998)
76. W. Lang, S. Shankar, J.M. Patel, A. Kalhan, Towards multi-tenant performance SLOs, in *IEEE 28th International Conference on Data Engineering (ICDE 2012)*, Washington, DC (Arlington, VA), 1–5 Apr 2012 (IEEE Computer Society, 2012), pp. 702–713
77. W. Leinberger, G. Karypis, V. Kumar, Multi-capacity bin packing algorithms with applications to job scheduling under multiple constraints, in *ICPP*, Aizu-Wakamatsu City, 1999, pp. 404–412
78. S. Listgarten, M.-A. Neimat, Modelling costs for a MM-DBMS, in *RTDB*, Newport Beach, 1996, pp. 72–78
79. M.E. Lübbecke, Column generation, in *Wiley Encyclopedia of Operations Research and Management Science*, ed. by J.J. Cochran et al. (Wiley, Hoboken, 2010)
80. N.R. Mahapatra, B. Venkatrao, The processor-memory bottleneck: problems and solutions. Crossroads **5**(3), 2 (1999)
81. S. Manegold, P.A. Boncz, M.L. Kersten, Generic database cost models for hierarchical memory systems, in *VLDB 2002, Proceedings of 28th International Conference on Very Large Data Bases*, Hong Kong, 20–23 Aug 2002 (Morgan Kaufmann, 2002), pp. 191–202
82. M. Mehta, D.J. DeWitt, Data placement in shared-nothing parallel database systems. VLDB J. **6**(1), 53–72 (1997)
83. S. Mehta, A. Neogi, ReCon: a tool to recommend dynamic server consolidation in multi-cluster data centers, in *IEEE/IFIP Network Operations and Management Symposium: Pervasive Management for Ubioquitous Networks and Services, NOMS 2008*, Salvador, 7–11 Apr 2008 (IEEE Computer Society, 2008), pp. 363–370
84. C. Meyer, P.A. Papakonstantinou, On the complexity of constructing golomb rulers. Discret. Appl. Math. **157**(4), 738–748 (2009)
85. U.F. Minhas, J. Yadav, A. Aboulnaga, K. Salem, Database systems on virtual machines: how much do you lose? in *Proceedings of the 24th International Conference on Data Engineering Workshops, ICDE 2008*, Cancún, 7–12 Apr 2008 (IEEE Computer Society, 2008), pp. 35–41
86. J.J. Moré, D.C. Sorensen, Computing a trust region step. SIAM J. Sci. Stat. Comput. **4**, 553–572 (1983)
87. M. Myer, Rightnow architecture, in *Proceedings of the 12th International Workshop on High Performance Transaction Systems (HPTS)*, Asilomar Conference Grounds, Pacific Grove, 7–10 Oct 2007
88. J. Nielsen, *Usability Engineering* (Academic, Boston, 1993). ISBN:978-0-12-518405-2
89. M. Odersky, M. Zenger, Scalable component abstractions, in *Proceedings of the 20th Annual ACM SIGPLAN Conference on Object-Oriented Programming, Systems, Languages, and Applications, OOPSLA 2005*, San Diego, 16–20 Oct 2005 (ACM, 2005), pp. 41–57
90. E. O'Mahony, E. Hebrard, A. Holland, C. Nugent, B. O'Sullivan, Using case-based reasoning in an algorithm portfolio for constraint solving, in *Proceedings of the Nineteenth Irish Conference on Artificial Intelligence and Cognitive Science (AICS)*, Cork, Ireland, 27–28th Aug 2008
91. P.E. O'Neil, E.J. O'Neil, X. Chen, S. Revilak, The star schema benchmark and augmented fact table indexing, in *Performance Evaluation and Benchmarking, First TPC Technology Conference, TPCTC 2009*, Lyon, 24–28 Aug 2009, Revised Selected Papers (Springer, 2009), pp. 237–252

92. D.A. Patterson, G.A. Gibson, R.H. Katz, A case for redundant arrays of inexpensive disks (RAID), in *Proceedings of the 1988 ACM SIGMOD International Conference on Management of Data*, Chicago, 1–3 June 1988 (ACM, 1988), pp. 109–116
93. H. Plattner, A common database approach for OLTP and OLAP using an in-memory column database, in *Proceedings of the ACM SIGMOD International Conference on Management of Data, SIGMOD 2009*, Providence, 29 June–2 July 2009 (ACM, 2009), pp. 1–2
94. H. Plattner, SanssouciDB: an in-memory database for processing enterprise workloads, in *Datenbanksysteme für Business, Technologie und Web (BTW), 14. Fachtagung des GI-Fachbereichs "Datenbanken und Informationssysteme" (DBIS)*, 2.-4.3.2011 in Kaiserslautern (GI, 2011), pp. 2–21
95. R.A. Popa, C.M.S. Redfield, N. Zeldovich, H. Balakrishnan, CryptDB: protecting confidentiality with encrypted query processing, in *Proceedings of the 23rd ACM Symposium on Operating Systems Principles 2011, SOSP 2011*, Cascais, 23–26 Oct 2011 (ACM, 2011), pp. 85–100
96. Y. Saito, C.T. Karamanolis, M. Karlsson, M. Mahalingam, Taming aggressive replication in the pangaea wide-area file system, in *5th Symposium on Operating System Design and Implementation (OSDI 2002)*, Boston, 9–11 Dec 2002 (USENIX Association, 2002)
97. SAP, BusinessObjects BI OnDemand (2012), http://www.biondemand.com/businessintelligence (Online). Accessed 13 Dec 2012
98. J. Schaffner, A. Bog, J. Krüger, A. Zeier, A hybrid row-column OLTP database architecture for operational reporting, in *Informal Proceedings of the Second International Workshop on Business Intelligence for the Real-Time Enterprise, BIRTE 2008, in Conjunction with VLDB'08*, Auckland, 24 Aug 2008
99. J. Schaffner, B. Eckart, D. Jacobs, C. Schwarz, H. Plattner, A. Zeier, Predicting in-memory database performance for automating cluster management tasks, in *Proceedings of the 27th International Conference on Data Engineering, ICDE 2011*, Hannover, 11–16 Apr 2011 (IEEE Computer Society, 2011), pp. 1264–1275
100. J. Schaffner, B. Eckart, C. Schwarz, J. Brunnert, D. Jacobs, A. Zeier, H. Plattner, Simulating multi-tenant OLAP database clusters, in *Datenbanksysteme für Business, Technologie und Web (BTW), 14. Fachtagung des GI-Fachbereichs "Datenbanken und Informationssysteme" (DBIS)*, 2.-4.3.2011 in Kaiserslautern (GI, 2011), pp. 410–429
101. S. Seltzsam, D. Gmach, S. Krompass, A. Kemper, AutoGlobe: an automatic administration concept for service-oriented database applications, in *Proceedings of the 22nd International Conference on Data Engineering, ICDE 2006*, Atlanta, 3–8 Apr 2006 (IEEE Computer Society, 2006), p. 90
102. J. Shawe-Taylor, N. Cristianini, *Kernel Methods for Pattern Analysis* (Cambridge University Press, Cambridge/New York 2004). ISBN:978-0-521-81397-6
103. V. Sikka, F. Färber, W. Lehner, S.K. Cha, T. Peh, C. Bornhövd, Efficient transaction processing in SAP HANA database: the end of a column store myth, in *Proceedings of the ACM SIGMOD International Conference on Management of Data, SIGMOD 2012*, Scottsdale, 20–24 May 2012 (ACM, 2012), pp. 731–742
104. S. Sivasubramanian, G. Pierre, M. van Steen, G. Alonso, Analysis of caching and replication strategies for web applications. IEEE Internet Comput. **11**(1), 60–66 (2007)
105. G. Soundararajan, C. Amza, A. Goel, Database replication policies for dynamic content applications, in *Proceedings of the 2006 EuroSys Conference*, Leuven, 18–21 Apr 2006 (ACM, 2006), pp. 89–102
106. B. Speitkamp, M. Bichler, A mathematical programming approach for server consolidation problems in virtualized data centers. IEEE Trans. Serv. Comput. **3**(4), 266–278 (2010)
107. M. Stonebraker, D.J. Abadi, A. Batkin, X. Chen, M. Cherniack, M. Ferreira, E. Lau, A. Lin, S. Madden, E.J. O'Neil, P.E. O'Neil, A. Rasin, N. Tran, S.B. Zdonik, C-store: a column-oriented DBMS, in *Proceedings of the 31st International Conference on Very Large Data Bases*, Trondheim, 30 Aug–2 Sept 2005 (ACM, 2005), pp. 553–564
108. Sybase, Sybase IQ (2012), http://www.sybase.com/products/datawarehousing/sybaseiq (Online). Accessed 13 Dec 2012

109. M. Tamer Özsu, P. Valduriez, *Principles of Distributed Database Systems*, 3rd edn. (Springer, New York, 2011). ISBN:978-1-4419-8833-1
110. The Apache Software Foundation, Apache Zookeeper (2012), http://zookeeper.apache.org/ (Online). Accessed 13 Dec 2012
111. S. Tozer, T. Brecht, A. Aboulnaga, Q-Cop: avoiding bad query mixes to minimize client timeouts under heavy loads, in *Proceedings of the 26th International Conference on Data Engineering, ICDE 2010*, Long Beach, 1–6 Mar 2010 (IEEE Computer Society, 2010), pp. 397–408
112. Transaction Processing Performance Council (TPC), TPC benchmark C (2012), http://www.tpc.org/tpcc/spec/tpcc_current.pdf (Online). Accessed 15 Dec 2012
113. Transaction Processing Performance Council (TPC), TPC benchmark DS (2012), http://www.tpc.org/tpcds/spec/tpcds_1.1.0.pdf (Online). Accessed 17 Dec 2012
114. Transaction Processing Performance Council (TPC), TPC benchmark H (decision support) (2012), http://www.tpc.org/tpch/spec/tpch2.14.4.pdf (Online). Accessed 13 Dec 2012
115. Transaction Processing Performance Council (TPC), TPC benchmark W (Web Commerce) (2012), http://www.tpc.org/tpcw/spec/tpcw_V1.8.pdf (Online). Accessed 17 Dec 2012
116. B. Urgaonkar, P.J. Shenoy, T. Roscoe, Resource overbooking and application profiling in shared hosting platforms, in *5th Symposium on Operating System Design and Implementation (OSDI 2002)*, Boston, 9–11 Dec 2002 (USENIX Association, 2002)
117. J.M. Valério de Carvalho, Exact solution of bin-packing problems using column generation and branch-and-bound. Ann. Oper. Res. **86**, 629–659 (1999)
118. A. Verma, P. Ahuja, A. Neogi, pMapper: power and migration cost aware application placement in virtualized systems, in *Middleware 2008, ACM/IFIP/USENIX 9th International Middleware Conference*, Leuven, 1–5 Dec 2008. Proceedings (Springer, 2008), pp. 243–264
119. W. Vogels, Eventually consistent. ACM Queue **6**(6), 14–19 (2008)
120. A. Watanabe, H. Yokota, Adaptive overlapped declustering: a highly available data-placement method balancing access load and space utilization, in *Proceedings of the 21st International Conference on Data Engineering, ICDE 2005*, Tokyo, 5–8 Apr 2005 (IEEE Computer Society, 2005), pp. 828–839
121. C.D. Weissman, S. Bobrowski, The design of the force.com multitenant internet application development platform, in *Proceedings of the ACM SIGMOD International Conference on Management of Data, SIGMOD 2009*, Providence, 29 June–2 July 2009 (ACM, 2009), pp. 889–896
122. T. Wood, P.J. Shenoy, A. Venkataramani, M.S. Yousif, Sandpiper: black-box and gray-box resource management for virtual machines. Comput. Netw. **53**(17), 2923–2938 (2009)
123. J. Wust, J.-H. Boese, F. Renkes, S. Blessing, J. Krüger, H. Plattner, Efficient logging for enterprise workloads on column-oriented in-memory databases, in *21st ACM International Conference on Information and Knowledge Management, CIKM'12*, Maui, 29 Oct–02 Nov 2012 (ACM, 2012), pp. 2085–2089
124. L. Xu, F. Hutter, H.H. Hoos, K. Leyton-Brown, SATzilla: portfolio-based algorithm selection for SAT. J. Artif. Intell. Res. **32**, 565–606 (2008)
125. F. Yang, J. Shanmugasundaram, R. Yerneni, A scalable data platform for a large number of small applications, in *CIDR 2009, Fourth Biennial Conference on Innovative Data Systems Research*, Asilomar, 4–7 Jan 2009. Online Proceedings, 2009, www.cidrdb.org
126. W. Zheng, R. Bianchini, G.J. Janakiraman, J.R. Santos, Y. Turner, JustRunIt: experiment-based management of virtualized data centers, in *Proceedings of the 2009 Conference on USENIX Annual Technical Conference, USENIX'09* (USENIX Association, Berkeley 2009), pp. 18–18. http://dl.acm.org/citation.cfm?id=1855807.1855825
127. H. Zhu, P. Gu, J. Wang, Shifted declustering: a placement-ideal layout scheme for multi-way replication storage architecture, in *Proceedings of the 22nd Annual International Conference on Supercomputing, ICS 2008*, Island of Kos, 7–12 June 2008 (ACM, 2008), pp. 134–144

128. J. Zhu, B. Gao, Z.H. Wang, B. Reinwald, C. Guo, X. Li, W. Sun, A dynamic resource allocation algorithm for database-as-a-service, in *IEEE International Conference on Web Services, ICWS 2011*, Washington, DC, 4–9 July 2011 (IEEE Computer Society, 2011), pp. 564–571
129. M. Zukowski, P.A. Boncz, Vectorwise: beyond column stores. IEEE Data Eng. Bull. **35**(1), 21–27 (2012)
130. M. Zukowski, M. van de Wiel, P.A. Boncz, Vectorwise: a vectorized analytical DBMS, in *IEEE 28th International Conference on Data Engineering (ICDE 2012)*, Washington, DC, (Arlington, VA), 1–5 Apr 2012 (IEEE Computer Society, 2012), pp. 1349–1350

Printed by Publishers' Graphics LLC
MLSI130723.15.14.57